IMPLEMENTING
TWI

Creating and Managing
a Skills-Based Culture

IMPLEMENTING
TWI

Creating and Managing
a Skills-Based Culture

Patrick Graupp
Robert J. Wrona

CRC Press
Taylor & Francis Group
Boca Raton   London   New York

CRC Press is an imprint of the
Taylor & Francis Group, an **informa** business

A PRODUCTIVITY PRESS BOOK

Productivity Press
Taylor & Francis Group
270 Madison Avenue
New York, NY 10016

© 2011 by Taylor and Francis Group, LLC
Productivity Press is an imprint of Taylor & Francis Group, an Informa business

No claim to original U.S. Government works

Printed in the United States of America on acid-free paper
10 9 8 7 6 5 4 3 2 1

International Standard Book Number: 978-1-4398-2596-9 (Hardback)

Library of Congress Cataloging-in-Publication Data

Graupp, Patrick.
  Implementing TWI : creating and managing a skills-based culture / Patrick Graupp and Robert J. Wrona.
    p. cm.
  Includes bibliographical references and index.
  ISBN 978-1-4398-2596-9
  1. Employees--Training of--United States. 2. Occupational training--United States. I. Wrona, Robert J. II. Title.

HF5549.5.T7G567 2011
658.3'124--dc22   2010022031

Visit the Taylor & Francis Web site at
http://www.taylorandfrancis.com

and the Productivity Press Web site at
http://www.productivitypress.com

Dedication

To our wives Arden Graupp and Inez Wrona

Without their patience, encouragement, and support
we would not have been able to maintain our focus on
bringing TWI back into the mainstream of training.

Contents

SECTION III TWI AND CULTURE

SECTION V EXPANDING TWI BY LEVERAGING JR, JI, AND JM

Foreword

Companies are under ever increasing pressure to remain on the competitive cutting edge. A fully engaged workforce is essential for doing this successfully, and Training Within Industry (TWI) is a powerful approach to creating and maintaining this engagement.

The increasing competitive pressure comes from a variety of forces.

- Economic development in Asia, South America, and Eastern Europe has increased the number of potential rivals challenging for customers loyalties.
- Incredible improvements in communication and transportation have converted those potential rivals in actual ones.
- Scientific and technological advances compress the half-life on any market offering's viability, increasing the demand for ever faster improvement and innovation in development, design, production, and delivery.

These three forces are significant under any circumstances. Add to this the fallout of the world wide economic recession the last few years. All organizations have to rapidly reconfigure how the bring value to market as customers have become more circumspect in terms of how they are going to satisfy needs that have changed in significant, discontinuous ways.

That workforce engagement is essential is also without question. A naive view might be that increased technological sophistication has increased the capacity for a select brain trust to do the hard 'thinking' of what to sell and how to make it, leaving the remainder of the organization to do nothing more than be button pushing monkeys for automated equipment and processes.

This belief that mind and muscle are separable is simply wrong, at least as wrong now as it was in the past. In the heyday of scientific management, it was thought that a select few could do the time motion studies to reveal how the remaining masses could work most effectively and efficiently. This missed incorporating in work design the subtleties of circumstances known only by those involved in actually doing work, and it missed incorporating the additional critical perspective other people might have brought to the design. Separating mind from muscle had no basis other than the elitist social construct of the day.

The need for broad engagement has only gone up, not down. Walk into any work environment—manufacturing, healthcare, and any service sector—and the number of distinct professions needed to accomplish work has gone up by many multiples and the sophistication of the equipment people use to complete their work has increased exponentially as well. Creating value is ever more a team effort, with the skills required of individual team members ever more challenging in their acquisition and demanding in their expression. Manufacturing is no longer the physically hard work of wrench turning—like Charlie Chaplin in Modern Times or a character in Diego Rivera's fresco, Detroit Industry. It is cutting edge physics, chemistry, and increasingly biology brought to bear in creating products useful to society. The amount of required know how is considerable.

Which brings us to training. Success depends on staying ahead of the curve, and staying ahead of the curve depends on engaging the minds and the muscle of everyone in the organization. How then, to get that engagement? It is unrealistic to expect that people will arrive with the skills already intact. Even were they products of the most successful education, and we know not all education is so successful, they'll lack the job-specific skills and knowledge to succeed. People could acquire skills through experience alone, but that is both time consuming and unreliable. How do you ensure people get the right experiences at the right time? Or, people could be developed in a mentored apprenticeship fashion. But that too takes considerable time and produces uneven results.

Therefore, an essential ingredient in being competitive is having a reliable system for developing skills. This is where Training Within Industry comes in: Job Instruction training to bring novices up to speed, Job Methods training so they can be active agents in improving what they did and how they did it, and Job Relations training that teaches the foundations of positive employee relations.

Graupp and Wrona bring many examples of companies that embraced these elements of TWI, improving their competitiveness by improving their capacity to fully engage their workforce productively. These examples can serve as inspiration and models for years.

With best wishes for continued success,

Steven Spear
Sr. Lecturer, MIT Sloan School of Management
Author, The High Velocity Edge—How Market Leaders
Leverage Operational Excellence to Beat the Competition.

Acknowledgments

Our objective for writing *The TWI Workbook* published in 2006 was to document the standardized "J" program methodology that, when delivered by properly trained trainers would enable organizations to attain the desired results from this training. In that book we acknowledged "all the people too numerous to mention who brought us into their plants and firms across the country *(and now the world)* to introduce their people to TWI." Although we find ourselves in a similar situation with this follow-on book, here we would like to personally recognize all the people that work for the companies that contributed to the case studies in this book and give credit where credit is due.

However, there is also group of people at these companies that contributed a significant amount of personal effort to make sure their story was accurately documented on behalf of their organization. We are indebted to Martha Purrier and Linda Hebish at the Virginia Mason Medical Center, Suzanne Smith at W. L. Gore, and Sam Wagner at Donnelly Custom Manufacturing for the enthusiasm they brought to this challenge. A special thank you to Mark Bechteler and Dean Burrows at Nixon Gear; Scott Curtis, Scott Laundry, Jamie Smith, and Jennifer Pickert at the Albany International Monofilament Plant, and Alan Gross, Dustin Dreese, and Jennifer Dietter at Currier Plastics. These people not only took time from their busy schedules to work wih us on their case study, they also opened their doors for hundreds of visitors from around the world to tour their plants in the Syracuse, NY area where they learned firsthand from those involved about the level of commitment it takes to reap the benefits of TWI, and the on-going effort needed to continue down this path.

We are very fortunate to be working with talented and dedicated TWI Institute Master Trainers, Richard Abercrombie and Mike Braml, who spend weeks away from home traveling across the US and around the world to introduce TWI and to train trainers for organizations to internalize the training. These professionals are always prepared to introduce companies to the "J" programs and to coach them on how to get started down the right path. Their job was made that much easier thanks to Lynne Harding who provides logistical support for the TWI Institute.

Our ability to write about implementing TWI at the level of detail found in the case studies in this book would not have been possible had it not been for a select group of distinguised authors that publicized the relationship TWI has with the Toyota Production System and Lean. We thank Robert "Doc" Hall, Jim Huntzinger, Jeffrey Liker, David Meier, Alan Robinson, Dean Schroeder, John Shook, and Art Smalley who opened the doors for the resurgence of TWI.

Introduction: A Compelling Need for Skills Training

For a variety of reasons and across all facets of the workplace landscape, organizations are finding more and more that they can't find people who possess the skills they need to get the work done. According to the *2005 Skills Gap Report*, a survey of 8,000 NAM members conducted by the National Association of Manufacturers (NAM), the Manufacturing Institute's Center for Workforce Success, and Deloitte Consulting LLC, there is a critical shortage of skilled workers in both breadth and depth. The report projected that this shortage would hinder the ability of U.S. manufacturers to compete in a global economy. That time has certainly arrived. Therefore, we can no longer wait to take action with the tools we have, and one of these is Training Within Industry (TWI) that was created for just such a need. Three events are coming together to accelerate the skills shortage at an inopportune time for manufacturers already struggling with how to compete in a global economy.[*]

An Aging Workforce

Projected demographics for 2010 showed that the workforce in the age range of 45 to 54 will grow by 21%, as the baby boomer generation

---

[*] Terry Wiseman, *A Perfect Storm Hits the Skilled Workforce*, Plant Engineering, December 2008.

workforce (age range of 55 to 64) grows by 54%. This raises the question of how companies will find qualified replacements as the boomers retire from the workplace.

A Compromised Educational System

Up to 50% of current students drop out of school in a majority of school districts, and only one-third of all graduating U.S. students can perform at the twelfth-grade level. That is not a good indicator of the quality of the workforce that will be available to manufacturers in the future as the skill level requirements for production workers continue to increase. This situation makes it that much more important for companies to capture their "tribal knowledge," the tricks and knacks that people develop on their own over years of experience on the job. This is the type of critical information and acquired knowledge that these people will take with them unless it can be quickly passed on to new employees. Otherwise, new employees will have to learn it by repeating the same mistakes their predecessors did when they learned these jobs.

Lack of Technical Apprenticeships

Companies do not support apprenticeship programs as they did in the past. Even when they do, individuals are not signing up due in good part to the fact that parents and high school guidance counselors do not understand the opportunities that exist for skilled people in industry today. To this list of requirements we add *Lean Manufacturing* that is placing higher demands on production workers, teams, and their leaders. In today's workplace, these people are increasingly being asked to run more sophisticated machinery and to make daily decisions that have a direct effect on the profitability of their organizations.

2005 Skills Gap Report—The American Manufacturing Workforce

According to the NAM survey, the shortage of qualified employees exists in the following critical areas:*

- Eighty percent of manufacturers surveyed are experiencing an overall shortage of qualified workers that cuts across industry sectors.
- The shortage of qualified workers is most acute on the front line, where 90% report a moderate to severe shortage of *qualified skilled production employees*, including machinists, operators, craft workers, and technicians.
- Manufacturers face the additional challenge of poor skill levels among current employees, with 46% reporting inadequate problem-solving skills among employees.

The magnitude of these shortages is better understood when we look at how high the bar was set by the creators of this survey. The following definition of a *qualified skilled production employee* was provided as a reference for companies to respond to the survey questions.

> A skilled production employee is defined as being able to operate manufacturing equipment in more than one process; capable of recognizing process improvement opportunities; and have knowledge of manufacturing equipment and processes sufficient to understand and resolve moderately complex production issues, provide preventive maintenance, and make routine repairs. The skilled production worker can also apply advanced problem solving and analytical thinking skills to troubleshoot non-routine production issues.

---

* 2005 Skills Gap Report—A Survey of the American Manufacturing Workforce, NAM, Washington, DC, 2005.

If the above definition accurately describes your skilled production people, then one must ask how many of these people are over 55, how many of these people will go into retirement in the next few years, and how will you replace these people when they do retire?

The Shrinking U.S. Workforce

A majority of the more than eight hundred companies that participated in the NAM survey were small to midsize companies with fewer than five hundred employees. However, similar data were published in 2005 from a study of the strategies related to an aging workforce at three major multinational companies. These companies shared the same challenges of a shrinking workforce as the small and midsize manufacturing companies that made up the bulk of the NAM report companies.[*]

The immediacy of the challenge can be seen in the following projections for the year 2010:

- The overall rate of U.S. workforce growth was 30% in the 1970s and 12% in the 1990s through 2005, and is projected to no longer grow beginning in 2010.
- The number of U.S. workers forty-five to fifty-four will grow by 21% as the number of fifty-five- to sixty-four-year-olds grows by 52%.
- The number of U.S. workers thirty-five to forty-four years old will decline by 19%, while this pool of workers is also expected to shrink by 19% in the United Kingdom, 27% in Germany, and 9% in Italy.

Our shrinking workforce is also impacting healthcare, energy, and the public sector, where efforts are already being made to do more with less in order to overcome shortages of qualified people. The problem

---

[*] Lorrie Foster, Confronting the Global Brain Drain, *Knowledge Management Review*, November/December 2005.

has a far-reaching effect. "The U.S. Government is expecting to also feel the crunch: Within five years, half of the federal Chilean workforce (U.S. citizens working in Chile) will be eligible to retire by 2010."[*] As priorities shift to staff retention and transferring knowledge, the need for better skills in how to lead and to train people is that much more critical than it was in the past.

In the healthcare industry, where costs appear to be out of control, nurses have long been in short supply, as are technicians and associated hospital staff. According to a 2002 report by the American Hospital Association, the nursing shortage "reflects fundamental changes in population demographics, career expectations, work attitudes and worker dissatisfaction."[†] According to this research, hospital nursing vacancies will reach eight hundred thousand, or 29%, by 2020. The number of nurses is expected to grow by only 6%, while demand for nursing care increases by 40%.

Once again, we run into the issue of a shrinking workforce. Women ages 35 to 49 constitute the largest number of RNs in the workforce, while RNs under the age of 35 constitute the smallest. One way to reduce this demand for nurses from a smaller workforce might be to train other staff members on how to perform administrative and other tasks that can be passed down to others so that nurses can devote more of their time to the patient care for which they are trained.

Impact of a Shrinking Workforce on Manufacturing

The results of this survey indicate that global competition will force U.S. manufacturers, in order to stay in business, to compete less on cost than on productivity, flexibility, quality, and responsiveness to customer needs. This shift in thinking is already pushing manufacturers

---

[*] Ibid.
[†] M. W. Stanton and M. K. Rutherford, Hospital Nurse Staffing and Quality of Care, Agency for Healthcare and Quality, Rockville, MD, 2004.

and other industries to put a premium on the skills, morale, and well-being of their people. Companies that do not treat people as their most important asset are running the risk of losing their best people to other companies that will be willing to pay a premium for highly qualified employees who possess skills that are in short supply.

In 2008, the first members of the baby boomer generation turned 62, the average age of retirement in North America, Europe, and Asia, where 80% of workplace growth will occur among people 50 years or older over the next fifteen years. It is projected that by 2050, 40% of Europe's population and 60% of its working age population will be people over 60, at which time Germany, Italy, Spain, and Japan could face economic crisis from having mounting pension obligations funded by shrinking workforces in these countries.[*]

These projections do not reflect the impact of all the forced retirements that resulted from the economic crisis caused by the financial industry meltdown of 2009, which is making a bad situation even worse. One can't predict the impact of the skills drain going on in the automobile industry caused by the early retirement of so many people, many of whom were not physically, mentally, or financially prepared to exit the workforce.

As the competition increases for a shrinking number of skilled people, organizations must also rethink the ways they manage, train, and treat their key people, especially their supervisory group. When layoffs and reduced hiring are the order of the day, these are the people companies must invest in to keep pace with the demands of the times, just like the properly trained supervisors who are credited for the successful expansion of U.S. industry during World War II.[†]

---

[*] It's 2008: Do You Know Where Your Talent Is? Why Acquisitions and Retention Strategies Don't Work, A Deloitte Research Study, New York, 2004.

[†] Alan G. Robinson and Dean M. Schroeder, Training, Continuous Improvement, and Human Relations: The U.S. TWI Programs and the Japanese Management Style, *California Management Review*. Volume 35.

The Disengaged Employee

According to Deloitte Research, downsizing, increased employer demands, job disenchantment, and technologies that keep people plugged into their jobs without human interaction have taken their toll. The only thing worse for employers than a shortage of skilled people is to have employees who would rather not be working for the company but make no effort to leave.

According to current surveys, more than half the workforce is fed up:

> Pollster Gallup has found that 80 percent of British workers lack commitment to their jobs, with a quarter of those being "actively disengaged" from their workplaces. The situation is worse in France, where only 12 percent of workers in France were engaged in their work. In Singapore, 17 percent of work force is actively disengaged, creating a corrosive force in organizations. Disenchanted workers pull down productivity, increase churn, and darken the morale of the people around them. The annual economic costs are huge: as much as 100 billion Euros in France, US$64 billion in the UK, US$6 billion in Singapore, and a whopping $350 billion in the U.S.[*]

The crucial source of this disengagement and the number one reason why people leave a job is the poor relationship they have with their boss. This situation is typically overlooked and leads to myriad troubles in the workplace. More than half the workforce in the U.S. was already fed up with their work as the country entered these tough economic times when jobs became scarce and turnover was no longer a true measure of employee commitment. With improvements in the economy underway, companies might increasingly be surprised to hear from employees what they have not heard in a while: "I quit."[†]

---

[*] It's 2008: Do You Know Where Your Talent Is? Why Acquisitions and Retention Strategies Don't Work, A Deloitte Research Study, New York, 2004, page 4.

[†] Joe Light, More Workers Start to Quit, *Wall Street Journal*, May 25, 2010, page D6.

Recruiters and human resource experts say the increase in employees giving notice is a product of two forces. First, the natural turnover of employees leaving to advance their careers didn't occur during the recession because jobs were scarce. Another factor making it harder for companies to retain employees is the effect of the heavy cost-cutting and downsizing during the downturn on workers' morale. A survey conducted last summer for the Conference Board, a management research organization, found that the drivers of the drop in job fulfillment included less satisfaction with wages and less interest in work. "Employees feel disengaged with their jobs, which is going to lead to a lot of churn as we come out of the recession," says Brett Good, a district president of Southern California for Robert Hall International, an executive recruiting firm.[*]

Companies would do well to make it a priority to properly train the people tasked with leading others if they expect to turn around their disenchanted employees. The best way to energize people is to provide them with the tools they need to get their jobs done in the most effective way. For the most part, people want to be proud of their work, and they want to be rewarded for the good job they do. The best way to channel that energy into positive workplace relationships is to provide supervisors with the skills they need to lead people just like TWI did during WWII, like Toyota has done since adopting TWI in 1951, and as you will learn from the case studies in this book.

Building Talent: The New Old Mindset

Redesigning jobs, improving working conditions, and ensuring that key people are properly trained and deployed go a long way in keeping people engaged. The most successful companies doing this are those that

[*] Ibid.

encourage their people to participate in improvement activities without the fear of losing their jobs as a result. Deloitte Research concluded that a different approach to training people was needed to both properly train them and to keep them engaged.[*] As you will see below, the three suggestions from Deloitte on how to better train had already been captured more than sixty years ago when the TWI program was created.

1. *Rather than push more information onto employees through conventional training, it is more important that they "learn how to learn."*

   Job Instruction training (JI) teaches supervisors how to quickly train employees to do a job correctly, safely, and conscientiously. The method emphasizes preparing the operator to learn by first finding out what they already know about the job and fitting the instruction into this existing mental framework. By also giving the reasons why the job in done a certain way, people can understand the importance of the technique and the negative ramification if done incorrectly. In these ways, the learner is engaged in the process of learning instead of just being told or shown how to do a job.

2. *The best way to develop critical talent is through the collaborative resolution of real-life issues ('action learning').*

   Job Methods Improvement (JM) is unique in its simplicity allowing supervisors and individuals to improve the way they do their jobs. The aim of the JM program is to help produce greater quantities of quality products in less time by making the best use of the manpower, machines, and materials now available. In using the JM method, supervisors are taught to work with others, especially operators, in coming up with new methods and to take on improvement opportunities in their immediate area of responsibility, ones that they can do something about.

---

[*] It's 2008: Do You Know Where Your Talent Is? Why Acquisitions and Retention Strategies Don't Work, A Deloitte Research Study, New York, 2004, pages 6 and 7.

3. *They learn not by pondering a hypothetical problem, but by directly tackling real issues.*

> In all of the TWI training modules, each participant learns by applying the method to an actual job from their own areas—the "learn by doing" approach. This is also true with Job Relations training (JR) which teaches supervisors how to build positive employee relations, increase cooperation and motivation, and effectively resolve conflicts. By bringing in actual "people problems" to class that require them to take action, supervisors learn to handle problems by gathering and weighing facts before taking action, and then to check results to evaluate whether the action taken has helped production. In each real life problem discussed in the class, trainees also consider how that problem could have been prevented.

Lean Manufacturing in the Western World: Underutilized People

The Toyota Production System focuses on the seven wastes of production (defects, overproduction, inventory, motion, processing, transportation, and waiting), to which *underutilized people* was added in the Lean production version of TPS when introduced in the Western world. As David Magee points out in his book, *How Toyota Became #1*:

> The secret to Toyota's success is … how it approaches its business as a whole, with an underlying focus on "respect for people." By maintaining a focus on one very lofty ideal, and by implementing and maintaining a business structure that encourages every employee to be actively engaged in pursuing the company's goals, Toyota is developing into a self-generating internally combustive enterprise.[*]

---

[*] David Magee, *How Toyota Became #1*, The Penguin Group, New York, 2007, p. 8.

This is what makes Toyota's model as applicable to a bank, a retailer, or a hospital as it is to a manufacturer. Regardless of the business you are in, workers in today's high-paced technological world simply must "bring their minds to work." They cannot check their brains at the door like in the past era of mass production, where long tedious production runs, changeover specialists, quality by inspection, etc., allowed them to mindlessly do their work, and get paid well for it. The old model in which the blue-collar workers were the muscle of the organization and the white-collar workers were the brains was shattered by the Lean reformation that took place in Japan after WWII and today's management is providing Lean training to give people the tools they need to function in the new model. In turn, Lean has also dramatically changed the roles of the supervisor who must replace his or her firefighting skills with leadership skills in order to solve problems and prevent fires from getting started in the first place.

Saving the Supervisor

> *To make the worker responsible for his job and for that of the work group is also the best—and maybe the only—way to restore the supervisor to health and function.*
>
> **Peter F. Drucker, Management, 1973**

"For a half century or more the first-line supervisor, especially in manufacturing and clerical work, has seen his role shrinking in status, in importance and in esteem. Where a supervisor was 'management' to the employees only a half-century ago, he now has, by and large, become a buffer between management, union, and workers. And like all buffers, his main function is to take blows."* Bob remembers well what

---

* Peter F. Drucker, *Management: Tasks, Responsibilities, Practices*, Harper & Row, Publishers, 1973, pages 279, 280.

it was like being a buffer between management and the union when he was a shop floor supervisor at the Chevrolet-Tonawanda General Motors plant in the late 1960s. He did not appreciate the value of that learning experience until, as an independent TQM manufacturing consultant, he learned that supervisors were no better trained for the job than he was twenty years earlier.

Supervisors throughout the world today are still separated from the people they supervise by a wall of resentment and suspicion that comes from their role as a "buffer" between management, union, and disengaged workers. The supervisor's role as a "firefighter" has not diminished with the introduction of Lean Manufacturing because management continues to pressure managers and supervisors to do more with less by instilling more discipline. Increased discipline may in fact be needed, but as Mike Rother cautions in his book *TOYOTA KATA*, greater emphasis should be placed on how people can sense and understand a situation so they can react to it on their own in a way that moves the organization forward.

"The thinking (of management) seems to be that if people in the organization would adhere more closely to their work standards and do what they were supposed to do, there would be fewer problems. Unfortunately it does not work this way. Keep in mind the second law of thermodynamics, or entropy, which states that even if we follow the work standard, a work process will tend to slip toward chaos if we leave it alone. No matter what, there will be problems that the operators, if left alone, will have to work around. The process will decay."*

The Role of Leaders at Toyota

In his book, Rother explains how Toyota manages from day-to-day embedding continuous improvement and adaptation into and across the organization. This is accomplished through the use of two *kata* (the

* *TOYOTA KATA: Managing People for Improvement, Adaptiveness, and Superior Results*, McGraw-Hill, 2010, pages 163 and 164.

Japanese word for behavior routines, habits, or patterns of thinking and conducting oneself) that are taught to all Toyota employees:

> The Improvement Kata—a repeating routine of establishing challenging target conditions, working step-by-step through obstacles, and always learning from the problems we encounter; and
>
> The Coaching Kata—a pattern of teaching the improvement *kata* to employees at every level to ensure that it motivates their ways of thinking and acting.

Although these behavior patterns are not visible, they are "a big part of what propels that company as an adaptive and continuously improving organization."[*]

"The primary task of Toyota's managers and leaders then does not revolve around improvement per se, but around increasing the improvement capability of people. That capability is what, in Toyota's view, strengthens the company. Toyota's managers and leaders develop people who in turn improve processes through the improvement *kata*. Developing the improvement capability of people at Toyota is not relegated to the human resources or training and development departments. It is part of every day's work in every area, and it is managers and supervisors who are expected to teach their people the improvement *kata*…part of how people are managed day to day."[†]

This brings us back to Peter Drucker who in 1973 wrote: "No organization can function well if its supervisory force does not function."

> The crisis of the supervisor would by itself be reason enough to think seriously about organization of the worker and working. For making the worker achieving, making him or her responsible, is the one way of making the supervisor effective as a resource for the worker and

---

[*] *TOYOTA KATA: Managing People for Improvement, Adaptiveness, and Superior Results*, McGraw-Hill, 2010, page xvi.

[†] Ibid, page 186.

the work group. Only by becoming a resource for knowledge, information, training, teaching standard setting, and guiding the achieving worker and his work group does the supervisor move away from the current untenable role of "supervising" people.[*]

Restoring the Supervisor to Health and Function

We have learned much from the history of TWI in the U.S. during World War II and from the Japanese experience with TWI after the war, when Japanese industry was running at less than 10 percent of its 1935 to 1937 levels. It was during this period of reconstruction that the three standardized TWI training programs for supervisors were adopted on a national level. Job Instruction training (JI) taught supervisors the importance of proper training and how to provide this training to their workforce. Job Methods training (JM) taught supervisors how to make the best use of the people, machines, and materials now available. Job Relations training (JR) taught supervisors how to get results through people by treating them as individuals.

The case studies in this book demonstrate how managers, supervisors, and team leaders benefit from the TWI training that provides the skills they need to become a resource for the worker and their work group. It is only when the worker and their work group accept responsibility for their work that the primary responsibility of the supervisor can be to create and maintain a good working environment by dealing with the day-to-day abnormalities in the worksite.

Our desire in writing this book is to help you understand how TWI can help you solve the compelling need for skills training for your people in the workplace. Throughout, we will show how companies have implemented TWI, following the same prescribed formulas so

---

[*] Peter F. Drucker, *Management: Tasks, Responsibilities, Practices*, Harper & Row, Publishers, 1973, pages 280, 281.

successfully used when the program was first developed during WWII, and how organizations are making TWI relevant in overcoming the challenges they face today. We believe that these great endeavors to "learn to do again what we did before" will build a strong foundation for future success at the worker and supervisor levels along the lines of companies in the U.S. during WWII, Toyota soon thereafter, and today the companies who so candidly shared their experiences with TWI in this book.

I

TWI TAKES HOLD IN THE U.S.—AGAIN

Chapter 1

Learning to Do Again What Was Already Done Before

> During war time, plants needed to use training in
> order to supply the needs of the armed forces. Now
> (with the end of the war), plants must use training
> if they are going to survive in competitive situations
> and if they are going to keep on providing jobs and
> wages for workers.*
>
> **—C. R. Dooley, April 1947**

Sixty years since the end of World War II, competitive situations around the world have taken their toll on companies that didn't heed this good advice from Dooley, who
served as director of the Training Within Industry (TWI)
program from 1940 to 1945. The need for good training,
as embodied by the TWI programs that were developed
to ensure victory in the war was clearly not only a wartime necessity. The global struggle would continue in a
different form even after the last shot was fired. These
same companies that abandoned the wartime training

---

* C. R. Dooley, *Report III: Vocational Training, I.L.O. (Montreal, 1946)*, Training
  Within Industry in the United States, Third Conference of American States Members of the International Labor Organization, Mexico City, April 1946, p. 161.

effort also ignored the early successes of organizations that ventured into Lean manufacturing before and during the 1940s, when wartime production needs forced thinking that was decades ahead of its time. Once the wartime imperative was over, they quickly fell in line with the mass production mindset that took over in the Western world with the end of the war.

In Japan, the story was quite different. The total destruction of the country's industry provided opportunities for the nation to think differently about how to make best use of their limited resources and a generation of workers lost to the war. Whatever their reasons, the Japanese embraced TWI and the program became an integral part of the revival of their production system that would soon become the envy of the entire world. A review of how Lean production evolved provides insight into how TWI became such an important part of that evolution that goes far beyond it being just a training program.

Toyota's Adoption and Dissemination of TWI

Our Lean voyage began like so many others when we read *The Machine That Changed the World.** Having revisited that book several times since learning the relationship TWI has with Lean production, we gained valuable knowledge from the thoughts of these learned authors on how Lean manufacturing techniques might be diffused around the world:

---

* James P. Womack, Daniel T. Jones, and Daniel Roos, *The Machine That Changed the World: The Story of Lean Production*, First Harper Perennial, New York, 1991.

We believe we have traveled farther and made more comparisons than anyone else.... So, where do we stand along the path to global diffusion in lean production? And what must happen for the whole world to embrace the system? Remember that as a practical matter there are only two ways for lean producers to diffuse across the world. The Japanese lean producers can spread it by building plants and taking over companies abroad, or the American and European mass-producers can adopt it on their own.[*]

Toyota has a fifty-year head start quietly diffusing TWI as an integral part of the Toyota Production System (TPS), and needless to say, there is much to learn by retracing the contributions TWI has had on Lean production training. It could be said, perhaps, that TWI came back to the United States when John Shook, author of *Learning to See*, was hired by Toyota and became a TWI instructor at Toyota's then new joint venture with GM in Fremont, California, New United Motor Manufacturing, Inc. (NUMMI), when it opened in 1984. With the completion of Toyota's first ground-up construction of a huge manufacturing complex in Georgetown, Kentucky, the TWI reintroduction continued. "Georgetown began production in 1986, and throughout the 1990s the plant routinely claimed the top spots in the widely watched J. D. Power and Associates quality survey for cars sold in the U.S."[†] Named North America's second best plant in 2001, second only to Toyota's Lexus plant in Canada, workers in North America were building Toyota cars "faster, better, and cheaper" than the cars coming out of Detroit. We

---

[*] Ibid., p. 240.
[†] Bumpy Road, *Wall Street Journal*, August 4, 2004, p. 1.

now know these North American workers were trained using TWI Job Instruction training.

The authors had an opportunity in 2005 to meet Jim White, who told us his "clock number" at Toyota's Georgetown plant was 14, which meant he was the fourteenth person hired at the plant. Jim was the training director in the early years of the plant and became a TWI Job Instruction trainer where he instructed the first supervisors of the plant how to teach jobs using TWI. Later, books like *The Toyota Way Fieldbook* and *Toyota Talent* by Jeffrey K. Liker and David Meier, also a former Toyota supervisor at Georgetown, would explain in detail how Toyota used the TWI JI method literally in the same form as they learned it in the 1950s. TWI is a foundational piece in the TPS system, so much so that it was put in place right up front when Toyota started the Georgetown plant.*

Marek Piatkowski was hired by TMMC in 1987 as the training manager for the Cambridge, Ontario, Canada, plant where he delivered the TPS Job Instruction program to train new hires as the plant staffed up for production. These are a few of his initial observations about Toyota's training style presented in his article "Training Recommendations for Implementing Lean," posted on the Lean Enterprise Institute website (www.lean.org) in 2008:

> Observation 1: One of my first discoveries about training at Toyota was there was very little written about TPS. There were no books or operating manuals. There were some brochures and

---

* In the current decade, with Toyota's rapid expansion worldwide, CEO Akio Toyota stated in response to massive quality recalls in early 2010: "Toyota's training of workers to maintain quality failed to keep up with the company's rapid growth" (Nikkei.com, March 17, 2010).

handouts, but nothing close to what we are used to, and there were no written policies defining what TPS was. Toyota was heavily dependent on the spoken word to train and sustain the knowledge of TPS from one generation to another.

Observation 4: Toyota requires five basic levels of knowledge and skills from a leader:

1. Knowledge of roles and responsibilities
2. Knowledge of job elements
3. Training skills
4. Leadership skills
5. Kaizen skills

In a conversation with Piatkowski, he told us that he was not aware in those days that what he had actually been trained to deliver by a senior mentor at Toyota was in fact the TWI JI program. It was not until he read articles on the history of TWI and books were published connecting Toyota to TWI that he learned where the TPS Job Instruction Program actually came from—the TWI JI training that was embraced by Toyota when introduced in Japan in 1951.

A Quick Review of the TWI Program

The TWI Service was one of the first emergency services established by the U.S. Government War Production Board "to help industry to help itself to get out more materials than had ever been thought possible, and at constantly accelerating speed." The TWI program was developed to assist defense industries to meet their manpower needs by training each worker to make the fullest use of his or her best skill up to the maximum of individual ability.* Industry responded by having its own people collect, develop, and standardize the TWI techniques as laid out by the TWI Service.

The challenge was to convince top management that standardized programs such as these could help them to meet the unique needs of their business and to neutralize the standard rejoinder that "our business is different." To achieve this, TWI devised an effective way of explaining the purpose of the J programs. It explained that supervisors have five basic needs:

■ Knowledge of the work
■ Knowledge of responsibilities
■ Skill in instructing
■ Skill in improving methods
■ Skill in leading

Knowledge of the work meant familiarity with the materials, machines, tools, processes, operations, and technical skills specific to the supervisor's industry. Knowledge of responsibilities involved an understanding of a company's specific situation, such as its rules, procedures, safety policies, interdepartmental relationships, and union contracts. TWI did not get involved in either of these two areas because each was unique and different for each company and industry. It did, however, directly address, through the three J courses, the need for skills in instructing, improving methods, and leading.†

---

* War Production Board, Bureau of Training, Training Within Industry Service, *The Training Within Industry Report: 1940–1945*, U.S. Government Printing Office, Washington, DC, 1945, p. 3.

† Alan G. Robinson and Dean M. Schroeder, Training, Continuous Improvement, and Human Relations: The U.S. TWI Programs and the Japanese Management Style, *California Management Review*, 35, 39, 1993.

1. *Job Instruction (JI) Training.* Trains supervisors how to instruct employees so they can quickly remember to do jobs correctly, safely, and conscientiously.
2. *Job Methods (JM) Training.* Trains supervisors how to improve job methods in order to produce greater quantities of quality products in less time by making the best use of the manpower, machines, and materials now available.
3. *Job Relations (JR) Training.* Trains supervisors how to lead people so that problems are prevented and gives them an analytical method to effectively resolve problems that do arise.

TWI trainers then trained people within industry who would, in turn, go on to train other people in industry, creating a multiplier effect that allowed a minimum of qualified trainers to reach a maximum number of people who could then respond to this challenge in the shortest period of time. To measure the impact of the TWI training on the war effort, the TWI Service monitored six hundred of its client companies from 1941 until it ceased operations in 1945. The last survey, performed just after TWI shut down field operations, detailed the following percentages of these firms that reported *at least 25% improvement in each of the following areas:*[*]

Increased production	86%
Reduced training time	100%
Reduced labor-hours	88%
Reduced scrap	55%
Reduced grievances	100%

[*] Ibid., p. 44.

Hidden within these percentages are recently acquired production figures that provide insight into claims that TWI shortened WWII by years:

> U.S. production totals in 1943 had included 86,000 planes, compared with barely 2,000 in 1939. Also 45,000 tanks, 98,000 bazookas, a million miles of communications wire, 18,000 new ships and craft, 648,000 trucks, nearly 6 million rifles, 26,000 mortars, and 61 million pairs of wool socks. Each day, another 71 million rounds of small-arms ammo spilled from U.S. munitions plants. In 1944 more of almost everything would be made.[*]

Also hidden in these percentages is what happened at these six hundred companies that were monitored by the TWI Service throughout the war. One of these was the Boeing Aircraft Company, where a program of intense training based on TWI was combined with an internally developed Lean production system to produce the B-17 bomber.

The Boeing Aircraft Company[†]

The B-17 bomber was already designed and tested before Boeing decided to build the plane at an existing facility south of Seattle that, although vacant and ideally located adjacent to an airport, was only half the size needed to assemble the aircraft with no land available for expansion. Boeing management accepted the challenge to overcome

---

[*] Rick Atkinson, *The Day of Battle: The War in Sicily and Italy, 1943–1944*, Henry Holt and Company, New York, 2007, p. 450.
[†] Bill V. Vogt and Robert "Doc" Hall, What You Can Do When You Have To: Parts I and II, *AME Target Magazine*, 15, no. 1, 1999.

these limitations by committing to having a high-morale, people-dependent system with teams, a strong suggestion system, and a program of intensive training as the foundation to building the most complex four-engine aircraft ever to be mass produced.

How all of the pieces came together is mostly lost to history, but there is enough evidence left of what they did to show how the creativity of people challenged by seemingly insurmountable constraints can rise above these obstacles. Long before anyone ever heard of Lean, the Boeing staff "invented" Lean concepts like flow by developing U-shaped final assembly lines that were fed by subassembly areas in other buildings. A takt time of seven minutes was measured for manual movement from station to station by using a clock they put up on the wall. Material handling was minimized by having subassembly areas feed directly to the line, and the use of dies was maximized by developing quick changeover techniques. The engineers had effectively created a Lean production system that changed the roles of the supervisor years before kaizen had the same impact in Japan.*

With production lines moving faster than at any time in the past, problems had to be solved as soon as they came up on the shop floor in order to keep the lines moving. The pace of production and frequency of engineering changes also required supervisors to be on the shop floor checking and coaching constantly. Since most of the experienced workers were sent off to war, supervisors, who themselves were most likely new at their responsibilities, had to lead a green workforce, making training and

---

* Masaaki Imai, *Gemba Kaizen*, McGraw-Hill, New York, 1997, Chap. 9.

knowledge of standard work instructions doubly important. Half of the new workforce of thirty-three thousand consisted of cowboys, farmers, fishermen, and lumberjacks. The other half was mostly area housewives.

This compelling need for supervisors to learn these essential skills was filled by TWI.

First, supervisors had to deal with a wide variety of people from different backgrounds who had never worked in a production environment before. TWI taught them to treat people as individuals and to get into problems early and prevent them where possible. The JR course case study of the first woman supervisor was indicative of problems faced at the time. Supervisors were taught on the job by TWI to break down common industrial tasks into easily digested, easily mastered steps for training new people. They also learned to cross-train employees to promote teamwork, and for people to learn how to take on broader responsibility within their work area. What is more, long before the Japanese word *kaizen* would be popularized, workers were trained with TWI to analyze and improve their methods to make them easier and safer to do. One such idea came from a "Rosie the Riveter" woman who, having trouble holding the heavy riveter over the wing she was working on, devised a two-jointed boom that would hold the weight of the machine even as she applied rivets across the entire surface of the wing.

The results of the TWI training, combined with their newfound Lean wisdom, created one of the many miracle stories to come out of the war. The production of B-17 bombers increased from 100 to 364 airplanes per month by March 1944, the peak production period. Production for that month numbered 15 fly-aways per day, 1 every 1.6 hours at a cost of $139,254, a 42.46% reduction in cost over a 32-month period accompanied by a 264% increase in shipments.[*]

Why Was TWI Dropped after WWII?

Industry and government viewed TWI as a wartime program when the TWI Service was shut down shortly after WWII ended. The United States was one of the few countries in the world where its industrial infrastructure was not damaged or destroyed by the war, enabling American manufacturers to shift to the mass production of consumer goods as quickly as they had shifted to the production of war materiel in 1940. Unlike Dooley, quoted at the beginning of this chapter, they did not see the application of TWI outside of the wartime production context, and with no competition from the rest of the world for their products, they didn't have any need to at the time.

The composition of the workforce changed dramatically as well as millions of people returned to their jobs from the military, displacing those millions of people who had been trained with TWI while they were gone. These men, who had returned to their country as heroes for winning the war, simply went back to their jobs and the way things were done before they left. And, being heroes, who was to tell them any differently? What is more, during the war TWI had provided grassroots attention to the concept of humanism in industry, which was

---

[*] Vogt and Hall, What You Can Do When You Have To.

allowed to flourish due to the unique circumstances of the makeup of the wartime workforce and the vital need to enlist the services of this group in a time of great need. But this style of leadership made management uncomfortable and, with the wartime needs behind them, they quickly reverted to the traditional command-and-control style of managing.

The situation in Japan, as fate would have it, was just the opposite. Having been utterly defeated in the war, the Japanese questioned their traditional systems and were open to learning from the Americans who occupied their country after the war. As with Toyota, all of Japanese industry embraced the TWI methods, along with the teachings of W. Edwards Deming and other programs introduced by the occupation. General MacArthur and his staff wanted to transform Japan into a successful capitalist society so they would never have to fight them again in a war. With programs like TWI in their arsenal, they succeeded beyond their wildest dreams.

Learning to Do It All Over Again

The TWI pilot projects conducted by the Central New York Technology Development Organization, Inc. (CNYTDO) in 2001 introduced the TWI program back into U.S. industry at a time when most companies were just getting started with Lean manufacturing.[*] With management perceiving Lean as a "house full of tools," supervisors learned to rely on scheduled events like 5S and Kaizen

---

[*] Patrick Graupp and Robert J. Wrona, *The TWI Workbook: Essential Skills for Supervisors,* Productivity Press, New York, 2006.

Blitzes to make even the smallest of changes in order to put out fires and meet their production numbers. Even when supervisors wanted to improve, they didn't have improvement skills, and it did not take long for people to view Lean as just another flavor of the month.

As was described in *The TWI Workbook*, "The first U.S. owned operation since the end of World War II to implement the TWI Program"* was the ESCO–Syracuse plant. ESCO is a world-class producer of precision casting parts for highly engineered products used in aircraft engines, power generation equipment, and missiles. Cellular manufacturing was introduced in 1995 to remove department silos, followed up soon after with the introduction of kaizen, synchronous manufacturing, and then Six Sigma. Although kaizen events were completed on almost all of the plant processes, there was still significant operator variation in the assembly of initial molds in the wax department. Since wax is the front end of the manufacturing processes, having to rework the molds increased touch time and cost. The on-time release of molds in the wax department was down to 73% for the year 2002, affecting both manufacturing cycle times and customer delivery.

"Senior management, after developing a Balanced Scorecard (BSC) and strategy map for its new strategy, had learned that the front end of the production process was a major opportunity to reduce rework and improve quality."† Paul Smith, director of human resources at the time, informed us that their buddy training technique

---

* Ibid., CD supplement, case study one, p. 2.
† Robert S. Kaplan and David P. Norton, *Strategy Maps,* Harvard Business School Publishing Corporation, Boston, 2004, p. 3.

of assigning one of the best assemblers to train new employees was not working. His team came to the TWI training in March 2002 in search of a training method that was both repeatable and verifiable, and they found what they were looking for in the Job Instruction training.

Discussions with Paul Smith resulted in Patrick setting aside three weeks to partner with CNYTDO by delivering the JI, JR, and JM programs at the ESCO plant located in Chittenango, New York. Patrick had been delivering TWI at Sanyo for many years, where they had done translation and layout work to the materials along the way to suit their needs. So he obtained copies of the original TWI trainer delivery manuals along with forms from the archives to be sure the training delivered to the ESCO people represented the original program without extraneous modifications that may have crept into the materials. In this remake of the manuals and materials, Patrick retained certain embellishments from the Japanese versions of the manuals that improved the delivery of the materials without changing the original content. He also updated the layout of the manuals using more modern desktop publishing software. Deliveries were spaced two months apart so Paul could put TWI to work upon completion of each module while Patrick created the new trainer delivery manuals for the next "J" program. Bob recreated training materials such as pocket cards and breakdown sheets for use by the new ESCO trainers.

CNYTDO also used this time to market TWI to its sister organizations in the Manufacturing Extension Partnership (MEP) network, which is a part of the National Institute of Standards and Technology (NIST). CNYTDO is one of hundreds of MEP locations across the United States that work directly with manufacturing companies to develop successful business strategies such as Lean manufacturing. The MEP network was positioned to quickly introduce TWI as part of its national Lean program to provide supervisors at small and midsize manufacturers with the skills needed to spearhead change. CNYTDO arranged with the MEP national office to conduct a well-attended web conference in April 2002 in which Bob introduced Training Within Industry as an already proven Lean training program.

The Rest of the Country Picks Up the Banner

Feedback from the web conference confirmed that companies across the United States were struggling to engage shop floor people in the Lean process. Randy Schwartz, director of the North Dakota MEP, was one of the first to understand the importance of TWI and make it part of the Lean program they were promoting. Attracted by the simplicity of the training and its history of success, Randy invited us to conduct TWI workshops he arranged for companies in North Dakota. A key part of these workshops was to get people interested in the JI training method by demonstrating poor training methods like telling alone and showing alone, just as is done in Session 1 of the JI class. These common failures demonstrate the need for a better way to train, and by training a person from the audience on how to tie the fire underwriters' knot using the JI four-step method, the deal is closed. Randy soon followed up by having one of his people, Jody Bock, become one of the first NIST MEP field persons trained to deliver the JR program. In her class was also David Palazzoli, field engineer for the Utica, New York, MEP (MVATC), who went on to be the first MEP field person to become a trainer for all three J programs.

Presentations like the one made in North Dakota soon spread to conferences and company meetings across the country. People, such as one process engineer from W. L. Gore in Flagstaff, Arizona, saw the presentation at an AME conference and took it back to their company, where they began implementing it on their own before they started more formalized training (see Chapter 8). At a Baxter Healthcare plant where they make drug delivery systems, several engineers had taken training with Toyota in Kentucky, where they first heard about TWI. They invited us to give the presentation to their plant management in Cleveland, Mississippi, and soon thereafter signed up for classes that culminated in training six of their staff to become trainers of the JI course. An almost identical course of events took place at McCormick & Company in Baltimore, Maryland, when Rod Gordon, the VP of operations, found TWI on the Internet late one night as he was struggling to find solutions to implementing standard work in the Lean program he was implementing there.

In the meantime, other MEP centers that participated in webinars and conferences began to see the application for TWI as part of the MEP Lean program. Conrad Soltero, a field engineer in the El Paso offices of the Texas Manufacturing Assistance Center (TMAC), learned about TWI from his own study and research. He led the way in introducing TWI in Texas, and was soon followed by many Texas MEP trainers, starting with Mark Sessumes, who became the TWI trainer for the Dallas office. As the number of trainers grew, TMAC recruited Patrice Boutier to lead the effort as TWI became a regular part of their Lean development program across Texas.

In Minnesota, Rick Kvasager of Enterprise Minnesota began a project at a client company in Alexandria, Donnelly Custom Manufacturing Company (see Chapter 5), that also led to his MEP having one of their field engineers, Mike Braml, become a TWI trainer to deliver TWI across Minnesota, including the school system. Susan Janus of the Massachusetts MEP and Tony Manorek of NEPIRC, the Pennsylvania MEP, became strong TWI advocates. Word spread to other MEP centers and they had the TWI Institute train trainers for them to deliver TWI locally. As of this writing, there are now TWI trainers at MEP centers in Alabama, Alaska, Arkansas, Colorado, Georgia, Idaho, Illinois, Indiana, Iowa, Massachusetts, Missouri, New Hampshire, New Mexico, New York, Ohio, Oregon, South Carolina, and Washington.

While the MEP centers cater almost exclusively to small and medium-size manufacturers, large companies began approaching CNYTDO for TWI training on their own and without any solicitation. CNYTDO then created the TWI Institute as a nonprofit subsidiary to service this sector and was soon providing TWI training to companies like HNI (the HON group of companies), ICI (owners of Glidden Paint), Northrop Grumman, Intel, Parker Hannifin, Terex, Raytheon, PPG, and Trane. The institute also did a project with the U.S. Postal Service. When we arrived to do TWI training at the shipyards of Northrop Grumman in Newport News, Virginia, where they make aircraft carriers, submarines, and other vessels for the U.S. Navy and Coast Guard, and have done so more than one hundred years, they showed us copies of the original documentation signed by the shipyard management at the beginning of WWII committing themselves to deliver the TWI program. We had, at that moment, truly come full circle with TWI.

Even more interesting, and we didn't learn this until much later, was that TWI had been used in healthcare and was delivered at hospitals during WWII when they were dealing with shortages of doctors and nurses due to the war. This was not unlike the situation in wartime manufacturing, and as it turns out, healthcare organizations and workers today are benefitting from Lean and TWI just as much as their manufacturing cousins. At Virginia Mason Medical Center, where they have studied Toyota and created their own Lean program, which they call the Virginia Mason Production System, they are energetically applying the TWI methods to ensure the success of their Lean initiatives (see Chapter 9).

Overseas, as well, interest in TWI has been brewing and companies in Canada, Mexico, Australia, Europe, Central and South America, and Asia have been actively pursuing the TWI methodologies. A large German auto parts manufacturer, once it achieved success at enforcing standard work using JI, had its plants across the world begin adopting TWI as a standard method of training. General Cable, a former subsidiary of Phelps Dodge, gathered trainers from plants in Brazil, Venezuela, Honduras, Costa Rica, and Peru to learn JI and become trainers. Some of these trainers later went and taught JI classes at GC plants in Thailand, New Zealand, and South Africa. In Mexico, Alfredo Hernandez was managing one of the five Thomas & Betts plants in Monterrey when he led a team that integrated TWI into their Lean system. He then began implementing TWI with the other plants there by developing two trainers each in JI, JR, and JM and sharing these training resources between the five plants. Their American parent company was so impressed with the results that they decided to implement TWI in their U.S. and Canadian plants as well. The LEGO Company in 2010 began a project using JI to standardize processes across three key plants in Denmark, Mexico, and Hungary.

Learning from Toyota

As we can see from the vast number and diversity of companies—in manufacturing, service, healthcare, and government—using TWI today, the program developed during the crisis of WWII had such strong foundational roots that it still gives the same astounding results in our current environment as it did when it was created. As we will discuss throughout this book and in the many case studies of companies who have successfully implemented TWI into their company systems, the principles upon which the TWI methods are based transcend time and culture. However, this is not something that everyone recognizes up front. There is a reason why TWI fell into obscurity in this country and why it took so long to rediscover.

When companies began to practice Lean, as we noted at the beginning of the chapter, they saw Lean as a toolbox full of tools that would fix things. They thought that if they just studied these tools and applied them energetically and correctly, then their problems would go away and they would improve and show exceptional results. They failed to heed, though, that easily ignored piece of advice that told them, in Lean, first you have to change the culture before you change the practices. They did not work on the people, who, after all, embody the culture of an organization, and ended up frustrated even after having expended much effort and expense learning and applying the tools of Lean.

In almost every case where we have gone into an organization to give TWI training, the reason given for pursuing TWI was that the company was having trouble sustaining the gains of their Lean program. In particular, they were having difficulty understanding, much less achieving and maintaining, standard work. The publication of *The Toyota Way Fieldbook* in 2006, then, was a real cause for celebration in the growing TWI community because it provided a clear understanding of the role TWI JI training played in stabilizing processes and standardizing work as the foundation for improvement.[*] When *Toyota Talent* followed soon thereafter, the floodgates opened and Lean practitioners were able to see exactly how the JI method could provide them with a means of getting to standardized processes, so vital to the success of everything they did in Lean.

Now that we know that TWI is the tool we need, the question is how we use it to get to where we need to go. In this book, we will endeavor to show how companies have successfully implemented TWI so that, in the end, they get the promised results of Lean. By showing the connection between TWI and Lean practice, we'll learn how

---

[*] Jeffrey K. Liker and David Meier, *The Toyota Way Fieldbook*, The McGraw-Hill Companies, New York, 2006, Chap. 4 and Chap. 6.

TWI forms the foundation for these Lean tools and how to most effectively integrate TWI into a Lean strategy.

Please note that it is not the intent or ability of this book to explain or teach the details of the TWI methods—that is done in *The TWI Workbook*. While we will reference throughout the many aspects and nuances of the TWI methods, to understand how the methods work and how to use them, a thorough reading of *The TWI Workbook* will be required.

Before we begin, it is interesting to note the results of our first success story, ESCO, and how these results align very neatly with what *The Toyota Way Fieldbook* would later codify as standard work. Table 1.1 shows how the experience of ESCO in achieving standard work closely correlates with the Toyota process that would be laid out by author Jeffrey K. Liker a few years later.

Referring back to the Boeing example during WWII, the ESCO–Syracuse plant is a good example of going back to the basics of management support, Lean production system, and TWI training. It now seems clear why the TWI methods had been abandoned, to our great misfortune, after the war. Although ESCO credits the TWI JI program for much of its success, JI would have only had a minor impact had the company not done the upfront work with cellular manufacturing, kaizen, synchronous manufacturing, and Six Sigma. In other words, it was only after *not* getting satisfactory results with these Lean programs that the company identified a basic problem with training that led them to JI, stability, standardized work, and bottom-line results. Now that companies across the globe are diligently studying and applying

Table 1.1 Comparison of ESCO Case with *The Toyota Way Fieldbook*

The Toyota Way Fieldbook, 2006, p. 56	*ESCO–Syracuse, 2002–2004*
Standard work is lacking when "the amount of time it takes to perform a given process varies tremendously from person to person, across shifts, or over time."	Significant variability existed in techniques used by assemblers. Significant defects existed in the initial completed molds.
The first step in creating Lean processes is to achieve a basic level of process stability.	Month-to-month on-time release of molds went from 73% in 2002 to 98.6% in 2004.
The initial level of stability is generally defined as the capability to produce consistent results in some minimum percentage of the time.	Month-to-month on-time release of molds is now maintained at 98 to 99%.
A simpler indicator (of process stability) would be the ability to meet customer requirements with quality products the first time through, on time (again, 80% or better).	Reduced rework by 96%. Shipped 25% more product with the same number of people. Customer on-time delivery improved 80%.

Lean concepts, the need for good fundamental skills as represented by TWI has come back into clear view.

Doing What Needs to Be Done

We have revisited the ESCO–Syracuse case study in this chapter because, like Boeing in the 1940s, the people at this plant did what they had to do before it was too late. Patrick completed the initial JM training on September 21, 2001, just ten days after terrorists took down the twin towers of the World Trade Center. Who could have predicted that this catastrophic event could also have destroyed this plant in Chittenango, New York, that employed around 380 people? Already threatened by foreign competition, they lost about 50% of their volume soon after the attack on the World Trade Center.

In the aftermath of the attack, demand for the precision casting parts they produced for highly engineered products used in aircraft engines and power generation equipment literally disappeared overnight. It is the opinion of the authors that this plant may not have survived had management not already taken action to improve performance. The plant continued with their strategic plan using TWI and reducing rework by 96% between 2002 and 2004. Job Instruction training was vital during this period, as operators were reshuffled when employment was reduced almost 50%, and then again when people were recalled as business returned.

After our first case study on ESCO was written for *The TWI Workbook*, JI moved beyond the wax department into the metal end, coating, and foundry areas as TWI was incorporated into new employee training. Competency profiles began to be maintained along with a plant-wide training matrix for cross-functional training of all employees within flow lines and between departments. TWI trainers audited operators for compliance with standards, and operators were trained and retrained on a continuing basis through their team leaders.

Just as in the great war that gave birth to TWI, plants today continue to grow and flourish using these time-tested principles. As Jeffrey Liker pointed out with *The*

Toyota Way, the fundamentals don't change even as we use these basics to improve and prosper:

> A PARADOX OF THE TOYOTA WAY is that though it is continually improving and changing, the core concepts remain consistent. We are continuously learning new aspects of the process and seeing different applications in different situations. Yet as our understanding deepens, the "basics" continually resurface, guiding decisions and methods.[*]

---

[*] Liker and Meier, *The Toyota Way Fieldbook*, preface.

Chapter 2

Tenacious at Nixon Gear, Inc.

Nixon Gear, Inc. is one of many small manufacturers across the United States that quietly go about providing good jobs for the people in their community by manufacturing a profitable product that is always in demand. Located in Syracuse, New York, and employing fifty people, the company is an original equipment manufacturer (OEM) supplier of high-precision, high-speed gear sets used in the manufacture of a wide variety of industrial products around the world. The original company was founded in 1920, but our story begins in 1973 when Samuel R. Haines (Sam) joined his father's company, bringing with him a business plan to expand beyond their local market. With sales under $1 million, Gear Motions was incorporated with the vision of developing a regional network of companies whose unique specialties could be leveraged to better serve their customers' wide range of gearing needs. Over the years, ventures into aerospace and the plastic gear market were scaled back so that today Gear Motions is comprised of two companies: Nixon Gear in Syracuse and Oliver Gear in Buffalo, New York. This case study is focused on Nixon

Gear, which generated global sales of $9.6 million in 2008 and $6 million in 2009, a drop caused by the global economic recession.

Sam developed a strategy early on for his company to provide exceptional quality and service to compete in the niche market for precision gear sets. That strategy was to provide a financial commitment to having the latest technology and equipment available and to invest in his people so that the company would have a highly trained workforce capable of operating and maintaining that equipment. In keeping with this strategy, Sam provided employee training at all levels of the organization that included Total Quality Management (TQM), Statistical Process Control (SPC), Just-in-Time (JIT), value stream mapping, Lean overviews, problem solving, and team building. In order to make the training stick, management at the company sent up to two-thirds of their people to be trained at one time, making substantial investments in production downtime in return for future benefits to be gained from the training. People always came away from these classes with renewed excitement, and with every intention of applying the knowledge and skills acquired to improve their jobs and the performance of the company. But not unlike most companies that try to do all the right things, people at Nixon Gear did a few things after being trained and, before long, the work got in the way and things went back, for the most part, to the way they were before the training ever happened.

While they tried to manage ongoing improvement initiatives internally, the people at Nixon Gear were in truth more effective when outside consultants were brought in to help them improve quality and their work-place business processes. The company had just come off their high-water earnings mark in 2000 when the CNYTDO partnered with Patrick Graupp to reintroduce TWI into the U.S. market by delivering a condensed JM class. This initial pilot project was actually held at Nixon Gear in 2001, with several of their employees attending in addition to other members of the Syracuse manufacturing community.

Employed by Sanyo Electric at the time, Patrick had only two days of vacation time that he could devote to this project. This time constraint provided us early on with the first of many lessons learned as to why the creators of the TWI program standardized the delivery of the J programs into five two-hour segments conducted over five consecutive days. Mark Bechteler, who was the manufacturing production manager at the time, felt there was too much information for his people to grasp in the condensed two-day delivery format:

> It [JM] was interesting but I didn't grasp the training in context to what was intended. It just seemed too simple. I believed there had to be some correlation with complexity of the training and the results you would get. I thought it was a good tool to add to what skills we had, but I didn't see it as any better or worse than any other improvement tool we had. And, I definitely had no idea at all of how it fit with the JR and JI training at that time.

Though they didn't run with the TWI program after this initial introduction, the seed was planted in Mark's mind to be cultivated in the not too distant future when he became a staunch advocate for TWI at Nixon Gear. Except for a short stint at a foundry for two years after earning his AAS degree, Mark had worked in various capacities for Nixon Gear since 1983 where he led the way for Nixon Gear to become one of the first gear grinding companies to comply with International Organization for Standardization (ISO) standards in 1996.

Sam Haines embraced the U.S. quality movement in the 1980s by consistently providing his people with the best tools and training available for them to make their work more productive and personally satisfying. Consistent with his years of employee-centered practices, on November 17, 2005, he took the first step to turning his family-owned company into a 100% employee-owned company. As posted on the company's website, www.gearmotions.com, Sam believes that an employee stock ownership plan (ESOP) "will give employees the opportunity to contribute to the (company's) success *and* reap the benefits. Even more important to the employees and the community, the company stays right where it is."

Three reasons listed on the company website for his decision to make the company employee owned show his strategy for keeping the company viable and give testimony to point one of W. Edwards Deming's fourteen points for management to "create constancy of purpose toward improvement":

■ The ESOP provides a seamless and stable transition of ownership, and because the company has reinvested its profits over the years, it is very strong financially, making the transition much easier on the company.

- An ESOP is a way to give back to the people, the company's most valued asset—who helped build the company over the years.
- Lastly and most important, an ESOP gives the employee-owners a piece of mind for their future. If the company continues to do well, everyone shares in the success!

This succession plan also involved selecting a new leader to steer the company in Sam's place. It would have to be a person with the same vision of the company's future and an understanding of the ever increasing demands on manufacturing Sam challenged since joining the company in 1973. He found that person in Dean Burrows, who became president in 2008. More on that subject after we take a closer look at Nixon Gear's Lean journey.

The new ISO 9001:2008 quality standard became official on November 15, 2008, and just two months later, Nixon Gear became one of the first gear manufacturers in the United States to become registered to the new standard. In his current role as manager of quality and continuous improvement, Mark says, "The reason to get out ahead on this standard means getting a jump-start on the competition. We've been a 'process company' [the ISO 9001:2008 standard advocates adoption of a process approach] for many years, and this registration validates how mature and robust the quality system we use every day to build our high precision gears is already operating."

President Dean Burrows stated: "This registration demonstrates how far our operating system has come on the company's continuous improvement journey. We are able to implement change in our factory at a rate much faster than many of our ground gear competitors." In addition to its new ISO registration, Nixon Gear holds many company certifications and quality awards, including the UTC SUPPLIER GOLD, Certified Supplier Award, the first gear manufacturer in the country to win this award.

Nixon Gear has come a long way, and TWI has been one of the key ingredients to its continuing success. Let's take a closer look at that journey, a story similar to many stories of other companies of similar size across the United States as they learn to compete in a global market, and the road they took to get to where they are today.

The Challenge

At the end of the 1990s, management was fully aware that the competitive bar in their industry had been raised for the new century because of the complexity of the design of the products they were now able to manufacture. This was made possible only because the company had invested in the latest equipment as part of its long-term strategy. They also knew that having this equipment was only half of the solution. The supply of highly skilled people in central New York that would be needed to operate this kind of advanced equipment was tapped out—most of them had moved out of the area during the drastic downsizing of the region's industries in the 1970s and 1980s, and a new generation of skilled people had not been brought up to replace them.

As the new century got into full swing, TQM, Lean, and Six Sigma efforts that had been initially well deployed throughout the plant had gradually begun to fall back. This caused the management team to become weary of training that never seemed to result in lasting change. It wasn't uncommon for employees to say, "Not again!" or "Why do we have to keep doing this?" when it came to training. Whenever these questions got to Sam he would simply say, "Until we get it right."

In 2003, the company embarked on another Lean initiative that involved training in all of the tools of Lean. This time around, with the help of outside assistance, these tools were put into practice and measurable benefits were gained. The struggle to maintain these gains, though, returned like a virus shortly after the external consultant left. Sam decided that it was now time to put an end to the backsliding by having his people take ownership of their Lean strategy. Training was planned to have his people brush up on their Lean manufacturing skills with the goal for them to direct their own Lean efforts without relying on outside experts and to revitalize all the efforts they had made up until that time.

Caution! Lean Teams at Work

One of the highlights of this renewed Lean training was a full-day, learn-by-doing Lean simulation exercise in which over half of all the employees participated. They were inspired by the fact that they were able to gain a practical understanding of Lean and make good decisions based on that knowledge. "I couldn't believe the productivity improvements our group was able to accomplish by changing from a batch to a pull system," said one team member. Another learner said that "combining the hands-on simulation with the text book learning really made it clear that going Lean was no longer an option."

For the next six months, teams received continued training in the tools of Lean in order to improve everything from the flow of paperwork to and from the office, to reduction in setup times, lead times, and quick turnaround, to major reorganizations of a number of work areas. Building on work begun the year before, one team was able to cut customer lead time by 33% and to get setup times reduced by as much as thirty minutes. A 5S team saved hundreds of hours previously spent looking for tools, while another team simplified the flow of paperwork and parts to triple the speed at which these parts traveled through the shop. Improvements continued at a slower pace going into 2004, as sustaining the gains emerged yet again to be the biggest challenge. This was now beginning to frustrate the workforce.

Determined to maintain the momentum and not let their hard-fought gains go to waste, the search for the next "right tool for the job" led Sam and Mark to visit ESCO Turbine Technologies–Syracuse, where a much publicized, employee-driven continuous improvement program had been successfully put into place. According to Mark, "We learned how the people at this ESCO plant were using TWI to improve training and drive standard work and continuous improvement—all things we knew we wanted to do ourselves but were struggling with." He continued:

Looking back to the original TWI training we participated in during 2001 and 2002, I recall us having the typical "we're different, it won't be so easy for us" type of skepticism. When we toured ESCO in 2004, we had very realistic expectations and reservations. While we still believed that we were different, we also saw a lot of similarities. What really struck us was how enthusiastic production employees were about the Job Instruction training and the level of understanding they now had about their jobs. We would not have believed how the simplicity of the JI training triggered a 76% reduction of rework in this one department had we not seen it for ourselves.

Resurrecting TWI at Nixon Gear in 2004

Having learned from the past the importance of getting people on board before moving forward on a Lean implementation, management decided to begin to reintroduce TWI by first scheduling two Job Relations training classes to assist in breaking down old barriers to change. By starting with the leadership piece, it was thought, they could build the skills to better manage the change process. Each class included employees from the shop floor, customer service, engineering and management—about 40% of all employees. The JR training served its purpose and loosened up the people, who were expecting more lectures, and they actually had fun because it was interactive and dealt with the real problems they were facing.

After the training, shop floor people were quick to point out to management that *they* were not using the JR four-step method on how to handle a problem:

1. Get the facts.
2. Weigh and decide.
3. Take action.
4. Check results.

They also encouraged management to practice the JR foundations for good relations that are based on treating all employees as individuals. By insisting that management "practice what they preach," the employees who were taught JR were actually able to force culture change both up and down the organization.

JI was introduced the following month to the same groups of employees who had completed the JR training previously. Manufacturing teams were then formed upon completion of the training, and the manager of manufacturing facilitated the creation of Job Instruction Breakdowns (JIBs) on how to train a person to operate each piece of key equipment in the workplace. Each team tasked with making the breakdowns had a mix of senior employees and newer, "green" employees. This combination bore fruit as the new people saw things differently from their more experienced colleagues and had new ideas on both how to do the jobs and how to make them better.

It was during these initial team activities that the value of having trained people first in the Job Relations four-step method proved its worth and kept team members focused on their objective to identify the current best way to do each job for the initial JIBs. As the teams began the work, they discovered that no two skilled operators in the plant did any one job the same way. The team members knew enough about Lean to understand that, although they were making good parts, their processes were riddled with variation, and this had to be remedied if they wanted to get the benefits of standardized work. The hard part would be convincing the people to follow the one best way for doing each job.

Needless to say, there was much debate over what the one best way should be for any job they were breaking down. But, much to everyone's surprise, the few instances when "who's right?" became a contentious issue were quickly settled when someone on the team would pull out the JR card to solve the problem. This allowed everyone to refocus on the objective and to look at all the facts in the job, including the opinions and feelings of the people who did them. By coming up with several possible actions, in other words, different ways to get the job done well, they were then able to evaluate and select the one best option that everyone could agree on. The teams then tried out the method, put it into action, and evaluated the results.

Using the leadership skills they learned in JR enabled the teams to stay on track by completing JIBs for 70% of their machines in the three months following the JI training. Mark was pleasantly surprised to see how operators willingly shared the knacks and tricks they had developed over time for doing their jobs. By capturing these as Key Points on the JIB, newer employees could now be trained in these knacks and tricks so that they no longer had to learn them on their own through trial and error.

As per plan, Job Methods was introduced several months after the JIBs were completed and the operators trained. In this way, JM could also be put right into use upon completion of the JM training. The JM training was probably the most exciting for employees because it gave them a chance to put their improvement ideas into use while breaking down jobs for improvement. A log was set up to track the ideas and the savings from the JM improvement proposals submitted, which were monitored continuously to measure the impact of the training.

We Did It—Again

With all three TWI modules now trained and under way, laminated JIBs were stored on the shop floor for supervisors to train new operators. People regularly pulled their yellow JR pocket cards out when problems arose, and JM proposals were being generated on a regular basis. With all of this TWI activity now progressing smoothly and energetically, management sat back and watched employees perform their magic. After a two-year lapse, the TWI training that had been restarted was finally beginning to show promise and provide the results it was designed to generate.

Looking back, Mark commented:

> We thought TWI (the first time we trained it) was so straightforward and user friendly that it would sustain itself for that very reason—once the training was completed the employees would run with it. This has probably been the case with most training at Nixon Gear, not just TWI. I would also say that management did not use or even fully understand the TWI method and principles and yet expected it to work at the floor level. And so within two years of the initial training, we weren't seeing the results we expected. Even so, it was clearly better than any

other program we had introduced in that it was still being used by the shop floor without management's urging, albeit infrequently, and without the enthusiasm we had at the beginning.

Now that they had the basic building blocks of TWI in place, in addition to reaping the big benefits TWI provided, they employed these basic skills to help reenergize their Lean program for which they had invested years of training and effort. The TWI skills added that final piece of needed expertise so that they could make things like standardized work, continuous improvement, and waste elimination really stick. In effect, they were able to now self-sustain their Lean effort without the need for constantly calling in consultants to point them in the right direction. With that, Sam felt that he had prepared the groundwork and was now ready for his final big initiative.

Incentive to Improve: A Fresh Start

Sam's succession plan was to pass the company on to those who built it—the employees—and he began the process for Nixon Gear to become an employee-owned company (ESOP). Sam knew that tapping into the collective wisdom of his workforce would be the key to the long-term success of the company. He had always managed in a highly participatory way, including employees in decision making and drawing upon employee knowledge for improvements. And he felt confident that the way forward for the company, built by his father, was to let these very people manage their own destinies.

A key part of this plan involved selecting his successor, a leader with vision and an understanding of the increasing demands of manufacturing in the competitive global marketplace. Dean Burrows was well prepared for the challenge of developing Nixon Gear into a global custom gear manufacturer. Dean had previously been vice president of operations for the Marietta Corporation and, before that, director of supply chain for Carrier Corporation's Residential Light Commercial International Division. With twenty years of broad experience from various industries, including three years overseas while living in France, Dean brought a fresh perspective on how to attain Sam's vision for the company they both shared.

Dean became the president of Nixon Gear in 2008. The new president brought with him the wisdom that, in order to get Lean fully implemented, the immediately visible physical changes that get people on board initially (i.e., 6S, new layouts, flow, etc.) only require a fraction of the effort. The real effort is in changing the attitudes of people to sustain those changes and to continuously improve upon them. This part takes time. Everyone talks about Lean being a journey, and according to David Mann, "The journey begins in earnest after the production floor has been rearranged.... Without a lean management system in place to support the new physical arrangements, people are left to rely on their old tricks for fooling the system, using familiar workarounds to get themselves out of trouble."*

---

* David Mann, *Creating a Lean Culture: Tools to Sustain Lean Conversions*, Productivity Press, New York, 2005, p. 5.

Knowing that the company had a history of not managing previous change efforts effectively, Dean challenged the management team right from the start by revisiting the existing company mission and values statements to identify and then question the core competencies of Nixon Gear. He asked questions like, "What is preventing us from being listed among the hundred best companies in the United States?"

Like so many other companies that limit their thinking to what they do, the people at Nixon Gear had set limitations by viewing themselves as a gear company. With Dean's prodding, they began to entertain the notion that they had, or could have, skills beyond that traditional view. Once they began to view themselves as a learning organization, questions arose as to what the business would be like if they were able to position the company so that it could adapt to other business opportunities. Attitudes began to change as people came up with ideas and actions that would be required to support that goal. It soon became clear to the team that Dean was right: "If we want to transform our business, we need to develop a culture of continual learning and continual improvement to do it."

In the process of restructuring the management team, they hired a new engineer to place more emphasis on activities driving change on the shop floor. To facilitate the change, Dean also reassigned Mark from the role of manufacturing manager to director of quality and continuous improvement and filled Mark's vacancy by promoting an experienced employee. In doing so, Dean conscientiously made continuous improvement one of the highest priorities of the company.

Mark already knew the role that TWI played in changing the culture, but he first had to get the new president on board with the program. Although Dean was not familiar with TWI, that changed quickly as he tagged along when Mark hosted visitors from other companies looking to learn about TWI by seeing it deployed at Nixon Gear. Dean read *The TWI Workbook* to learn the TWI fundamentals and *The Toyota Way Fieldbook*, where he learned how TWI could be implemented at Nixon Gear. After many discussions on why the company was not getting the full success Mark had anticipated from TWI, they agreed that the company should refocus its efforts by making TWI an integral part of their Lean strategy to standardize work and continuously improve.

Resurgence of TWI

Although TWI had been used throughout the plant for several years, Mark wasn't really sure why they weren't getting the full results he had originally hoped for. Things were different now because, in his new role, it was his responsibility to decide on what to do about it. Having completed all of the ten-hour classes for JI, JR, and JM, he decided to learn more about the TWI program by taking the forty-hour train-the-trainer programs to learn how to teach the ten-hour classes when they were conducted locally at the TWI Institute. He attended the Job Relations training in September 2008, and that gave him a whole new appreciation for TWI.

"I had a fresh perspective," Mark explained, "and was really excited to get back and implement what I had learned. Within the next few months I completed the JI and the JM train-the-trainer classes as well. These TWI trainer programs gave me a far deeper level of understanding of how the three disciplines actually do work together like a three-legged stool. I think the level of understanding attained makes the whole TWI program so much more powerful, and I would recommend any company to do the same."

Mark gives the following observations about what lessons were learned by going deeper, with his leadership, into the finer points of the TWI methods.

Job Relations Training

We originally used JR properly as the foundation for good working relations. It is the rules of play, and how we now resolve and prevent issues. It keeps us on task and keeps egos out of the room. What's different this time around is the "prevent." JR is great at resolving issues, but the real value we are now getting by using the other side of the JR yellow pocket card, the foundations for good relations, is to prevent problems. We missed this the first time around.

Job Instruction Training

We originally underestimated the value of JI and its connection with standard work. We did a good job creating Job Instruction Breakdowns (JIBs) and placing them at all of our work centers, but we failed in any formal follow-up. We now use JI as a tool to train the standard work. Then we audit the training. This is what we missed originally. We will use JI to train a process or method of doing a job. The way each job is consistently performed is standard work. We monitor and track productivity and, when we see variation, we audit the job using the JIB and then retrain the operator as required.

Job Methods Training

We introduce JM as processes are stabilized to improve how each job is performed. Operators now use the JM pocket card to improve unstable, unrepeatable processes by breaking each job down into details that can be eliminated, combined, rearranged, or simplified. Believe it or not, we missed this connection JM has with continuous improvement in 2004.

JR, JI, and JM

The new Nixon Gear model is really quite simple—JR, JI, JM—almost too simple, but we are implementing it and are seeing great results already. We gave our people the skills of JR to help them stay on task and to always focus on the objective. That opens the door for people to create process stability using JI and, once you have stability, improve the job using JM. Repeat. This simple continual loop is what we overlooked originally, and TWI makes it so easy to do. We also missed the mentality that it is not a one-time improvement activity. TWI is the tool for continuous improvement, not just one-time events. Each J program has its own special purpose, and can be used every day—not just for special activities. Our training will emphasize this as we make the TWI model part of our culture, not just a tool for a certain job, but every job.

Actualization of Lean/TWI: The Precision Compressor Gear Cell Story

With Mark in his new position, Dean set an initial goal for him to conduct two kaizen events each month to quickly get people reenergized around Lean. Having been down this road before, Mark knew he needed to start out with a quick success. He brought his carefully selected list of problem areas to review with Dean so they could determine the best place to attack first. Dean immediately set Mark's list aside and suggested that he begin with a kaizen event on the precision compressor gear cell. Mark was totally taken aback. This was the cell where they had already run numerous kaizen events, and it was now recognized throughout the organization as the model of efficiency in the shop. Why, he thought, should we go back to a location where all of the key improvements had already been implemented? And what would happen, he worried, if this initial event got stalled because the process was already running so smoothly?

The precision compressor gear cell was originally created in 2002, when an employee team, applying the tools of Lean, made better use of several machines that were being operated by four different people at various locations throughout the shop. The operators were producing high-volume compressor parts in batch lots of one hundred to two hundred pieces. This initial project had been very important to the future of the company for a number of reasons. One was that the customer for these parts expected annual price reductions to keep these parts at Nixon Gear. Another reason was that the five employees working on the team had an opportunity to apply what they learned by converting this traditional manufacturing batch process into a Lean process with smaller lot size and zero queues. According to Mark, this change was a real paradigm shift in thinking at Nixon Gear, where they were guilty of saying, "Sure Lean can work for high-volume manufacturers, but not here. Things are different at a job shop."

The flow of material and paperwork was improved so that these machines could be arranged into the precision compressor gear cell that resulted in one operator being able to triple the speed of parts through the shop while cutting inventory in half. Further improvements were made down the road when a new machine was integrated into the cell that continued to operate like clockwork. Mark had good reason to question Dean by asking, "What was there to gain?" "There is a method behind my madness," replied Dean as their meeting ended.

From Dean's perspective, this cell produced six different gear sets (see Figure 2.1), for a total of 15,000 sets per year, and this was just the place where the company's expertise could be mined. Gear grinding is a complex business due to the extremely tight tolerances involved, and the goal is always to improve contact ratio between gears to lower the noise in our cars and air conditioners and all the products used in our everyday lives. What is

Figure 2.1   Sample of gears made in the Precision Compressor Gear Cell. (Reprinted with permission of Nixon Gear, Inc., Syracuse, NY.)

more, precision in this area extended the gear set life, eliminating expensive repairs. Since the grinding of precision hardened parts was a core competency of Nixon Gear, Dean had decided even before talking with Mark to have him conduct the initial kaizen event on the precision compressor gear cell.

Learning to See—Again

Mark quickly formed a new precision compressor gear cell team that was made up of six volunteers: the cell operator from each of the two shifts, the person who operated the cell when it was first formed, a quality technician, the new manager of manufacturing, and himself, the director of quality and continuous improvement.

Tom Cartner, the original operator of the precision compressor gear cell, explained how the new team quickly discovered that, by mapping how the parts flowed and how the operator moved within the cell, the second shift operator was actually batching ten parts between operations instead of traveling with each part from machine to machine to maintain one-piece flow. This opened up Mark's eyes to what was really happening in the cell because "the production reports indicated that each operator was producing the same number of quality parts per shift, so there was no way to determine that one operator had introduced variation into the process unless a major quality issue occurred as a result of the change." For their part, the operators believed that "as long the numbers were right at the end of their shift, they were OK."

The reason why the operator was batching parts was that the job was physically very demanding. According to Joe Boehm, the day shift operator, "The large gear weighs around 24 pounds, so it wears on the operator when we have a steady run of six of these parts each hour through the cell for a total of forty-eight parts run on each shift. That's a lot of lifting (48 × 24 pounds = 1,152 pounds)! We also had to do a lot of turning and twisting of the body as we carried each part from machine to machine. We really got a workout, and I went home tired at the end of each shift."

Quality technician Jerry Tarolli, who was on the original team that turned this multioperator operation into a one-person process, informed us that "the original Job Instruction breakdowns were created to train the operator to operate each machine, one by one. We never thought to also create a JIB to train the operator how to run the cell as a continuous process. That was a big miss."

With these insights, the team set out to redo their JIBs and retrain in order to create standard work around the continuous flow of parts through the cell. This immediately set them up against the complaints of heavy lifting and twisting, which the operators on the team insisted were big issues that had to be resolved. These were issues missed in the original development of the cell and the improvements that had ensued since then. Using their Lean tools, they went about rearranging the location of the machines so that they could be connected by short conveyors that would transport the heavy gears instead of them being lifted by an operator twisting between the two machines. Placement of the parts was also enhanced to ease the transport from palette to machine. Before and after pictures of the process are shown in Figures 2.2 and 2.3.

Figure 2.2 **Precision Compressor Gear Cell before the kaizen event. (Reprinted with permission of Nixon Gear, Inc., Syracuse, NY.)**

Figure 2.3 **Precision Compressor Gear Cell after the kaizen event. (Reprinted with permission of Nixon Gear, Inc., Syracuse, NY.)**

Mark summed up the impact of starting their Lean/ TWI voyage with a kaizen event in their most productive cell by stating, "Everyone in the shop was energized by what was accomplished in the precision compressor gear cell, and Sam and Dean came through by providing the support we needed to keep things moving." Following this success, new teams were then formed to reorganize other department layouts to better facilitate product flow and to reduce lot sizes. Their 5S activities were revitalized as machines were cleaned, relocated, and painted, and carts were purchased to make it easier to move smaller lot sizes from location to location. Visuals that were once just talked about were soon fabricated and deployed throughout the shop. As all these changes were implemented and jobs were improved, new JIBs were made so that operators could be retrained to sustain the gains of the changes.

Now that Mark did not have the distraction of having to get product out the door, his new responsibility for continuous improvement led him to send his people back yet again to the precision compressor gear cell to see if they could further improve upon the new method they had just established. He had the people apply their TWI Job Methods skills this time, now that the standard was firmly established using JI, to see if even more improvements were hidden in the new process. Additional improvements were in fact found and implemented, and are combined and shown in Table 2.1, along with all of the other improvements generated by applying their Lean tools.

Table 2.1 Overall Impact on the Precision Compressor Gear Cell

Kaizen Metrics	Baseline	Target	Actual	% Improvement Actual/Baseline
Space (sq. ft.)	1,092	850	590	46%
Walking (ft.)	10,290	8,500	2,100	80%
Parts travel distance (ft.)	60	30	23	62%
Lead time (days)	2	2	2	N/C
Quality (PPM)	793	500	48	94%
Parts per hour	6	6	7	17%
Changeover (min.)	120	120	120	0
Ergonomic stressors	34	27	26	24%
5S rating	12.0	25	22	83%

Reprinted with permission of Nixon Gear, Inc., Syracuse, NY.

Moving Forward with JM in Other Departments

Fresh off a successful experience with JM in the precision compressor gear cell, employees quickly learned why they were taught in the ten-hour class to question every detail by asking *why, what, where, when, who,* and *how, in that order,* when doing a JM breakdown. The reason is that we don't want to waste our time finding a "better way" to do a detail, the answer to the question *how,* if it is not necessary to perform that detail at all, the answer to the question *why.* This insight was demonstrated when the second shift operators did a JM breakdown on their method to spin out grinding wheels on all machines at the end of their shift. This was a job they had been doing for as long as anyone can remember in order to remove oil from the wheels, which causes wheel imbalance, leading to quality issues for the day shift.

By just asking the first two questions of *why* it is necessary and *what* its purpose is, the employees discovered that this procedure could be eliminated because, unbeknownst to anyone in the plant, the newer machines that had been installed over the past several years automatically balanced the wheels electronically. This hidden gem provided a gain of thirty minutes per day of production time on four machines, for a total of two hours each day. The team was pleasantly surprised when they learned that this change, which they could put into effect immediately, would translate into an annual savings of $51,600. This is a good example of what Alan Robinson and Dean Schroeder discovered about Job Methods training when doing research for their article about TWI:

> JM is a course in continuous improvement. It not only indoctrinates people into an "improvement" frame of mind, but teaches them how to find opportunities for improvement, how to generate ideas to take advantage of these opportunities, and then how to get these ideas put into practice. The results of such training can be impressive, and are not hard to document.[*]

Mark's team found further opportunities for improvement when they applied JM in the grinding department on a pretty straightforward load/unload of a gear grinding operation. Their initial observation indicated that

---

[*] Alan G. Robinson and Dean M. Schroeder, Training, Continuous Improvement, and Human Relations: The U.S. TWI Programs and the Japanese Management Style, *California Management Review*, Winter 1993, p. 51.

the operator was utilizing a steady process with good product flow and very little waste. Creating a JM breakdown (see Figure 2.4), though, led the team to analyze where the part came from and where the part was sent before and after this processing. By documenting every detail, they learned that the part had to travel 30 feet across the shop, where it was stored while waiting to be de-burred. When parts built up at the burring station, they were loaded into a washer to remove the oil before de-burring. This was a significant discovery because the process of moving, storing, washing, storing, burring, and moving was done for about 75% of their products.

The Job Methods improvement proposal (see Figure 2.5) illustrates how simple it was for the team to justify building a prototype mobile burring station to eliminate twenty-four labor hours from the process, reducing lead time by twelve days. Parts found in the shop were used to build the prototype that was located in the grinding department for evaluation. The team expects to generate additional savings by fabricating two or three more mobile burring stations to completely eliminate the need to transport parts to a central burring station. This change illustrates the objective for Job Methods as found at the top of each JM pocket card: "A practical plan to help you produce *greater quantities* of *quality products* in *less time* by making the **best use** of the **Manpower, Machines, and Materials now available**."

The company has also changed their focus on how they now use JM when it comes to looking at cost savings from improvement. While in the past they would document cost savings as the primary objective, Mark says, "Now we use JM to identify waste—period. We look

Job Breakdown Sheet

Product: Flowserve 61-415-0056					Made by: Dave M.						Date: 9/30/09				
Operations: Turn/Hob/Broach/Burr					Department:										
	Present/Proposed Method Details	Remarks			Why-What	Where	When	Who	How	Ideas. Write them down; don't try to remember	Eliminate	Combine	Rearrange	Simplify
		Distance	Time/Tolerance/ Rejects/Safety											
1	Turn blanks													
2	Move to rack	30 ft.	Wait in queue		×					Move machines in same area to create a small cell	×		×	
3	Remove from rack and take to hobber	60 ft.				×				Combine all operations and eliminate queue			×	
4	Hob gear teeth													
5	Move to rack	60 ft.	Wait in queue		×	×					×			
6	Remove from rack and take to broach	200 ft.												
7	Broach keyway					×							×	
8	Move broached parts to rack	120 ft.	Wait in queue		×									
9	Remove from rack and take to burring	30 ft.				×					×		×	
10	De-burr parts													
11	Move to QA	20 ft.												

Figure 2.4   Job Methods breakdown sheet for Flowserve: Part 61-415-0056. (Reprinted with permission of Nixon Gear, Inc., Syracuse, NY.)

Improvement Proposal Sheet

Submitted to: Management team

Made by: Dave M.   **Department:**

Product/part: Flowserve 61-415-0056   **Date:** 09/30/09

Operations: HG84/HB01/Burr

The following are proposed improvements on the above operations:

1. Summary

Rearrange machines in area to accommodate single-piece flow, combining hob, broach, and burr operations.

2. Results

	Before Improvement	*After Improvement*
Production (one worker per day)	200 pieces	200 pieces
Machine use (one machine per day)	1	3
Reject rate		TDB
Number of operators	3	1
Other: Lead time	14 days	2 days

3. Content

Before		*After*	
Turn blank	12 hours	Turn blank	12 hours
Queue		Hob	13 hours
Hob	13 hours	Broach	6 hours
Queue		Burr	6 hours
Broach	6 hours		
Queue			
Burr	6 hours		
Manufacturing time:	37 hours	Manufacturing time: 13 hours (all operations internal to turning)	
Total time:	14 days	Total time: 2 days	

Projected savings: 12-day lead time reduction

24 hours labor reduction

Note: Explain exactly how this improvement was made. If necessary, attach present and proposed breakdown sheets, diagrams, and any other related items.

Figure 2.5   Job Methods improvement proposal for Flowserve: Part 61-415-0056. (Reprinted with permission of Nixon Gear, Inc., Syracuse, NY.)

for and eliminate waste." This approach has "opened our eyes" to all forms of waste because, oftentimes, improvements seem to have no perceived impact on cost. This was just the case for a JM breakdown in the precision compressor gear cell that identified the operator opening and closing a heavy door twenty-four times each hour to access the lathe. It cost the improvement team just $100 to install a new automatic on/off switch to eliminate the operator performing this task, though with no immediate cost savings impact. However, ergonomic stressors for the operator went from 42.5 to 0 on a job that was already physically demanding for operators handling the heavy 24-pound parts. The impact the change had on the quality of life for the operators running the cell was enormous. Basically, Mark believes that any waste they can remove from a process will, in the end, ultimately reduce cost.

The Follow-Up Process

Now that all three of the TWI programs were up and running and fast becoming a part of the culture of the company, Mark and his team went back to see how they could close the final loop by making sure they didn't fall back on their old ways, as was their experience so often in the past. He wanted to use TWI as a way of self-generating itself so that the methods themselves ensured their continued usage. Since Step 4 of each TWI method is a follow-up process, he had to be sure this important part of the program was being fully utilized.

Productivity boards are maintained at each machine for operators to track hourly output (takt). Inconsistent output indicates that the process is unstable. Consistent adherence to standards demonstrates process stability. When a process is not demonstrating stability, the first step the people at Nixon Gear now take is to use the Job Instruction breakdown to audit whether the operator is performing the job as trained. If operators are not performing the job as identified in the JIB, the next step is to verify that the JIB is still accurate. If it is determined that the JIB is not accurate, a new JIB is created and then used to retrain all of the operators. Regular audits are then performed to ensure that every operator is doing the job as trained, and this procedure continues until the process is stabilized as demonstrated by consistent output. See Figure 2.6 for an outline of Nixon Gear's continual improvement model.

Job Methods fits in extremely well with this continual improvement model. When operators see an opportunity for improvement, they are now trained to evaluate whether there is a better, easier, or faster way to do the job by doing a JM breakdown and practice rather than waiting for another Lean event. This is what the team that conducted the kaizen event in the precision compressor gear cell did when they revisited the amount of lifting and turning operators had to go through when handling the 24-pound castings in that cell. Though they had made substantial improvements to the productivity of the cell, by auditing the work and finding it was not being performed to standardized, they were able to identify the reasons for that variation and used JM to make the needed changes.

By following up consistently on whether jobs are adhering to standard work, as evidenced by the consistent output of quality parts from the processes, the people at Nixon Gear are able to immediately act on any variations.

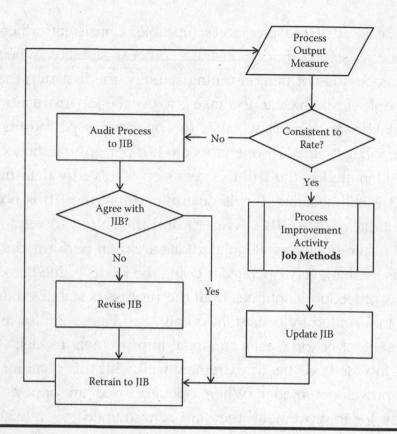

Figure 2.6 Continual improvement model using TWI. (Reprinted with permission of Nixon Gear, Inc., Syracuse, NY.)

This does not merely mean retraining operators on those standards, but first checking to be sure the correct training was done using the correct JIB, making adjustments to the JIB if needed, and then making improvements to those procedures when opportunities present themselves. Mark felt that, with this model, he had finally found the true secret to continuous improvement.

Conclusion

As mentioned at the beginning of this case study, the management of Nixon Gear initially thought TWI was so straightforward and user-friendly that employees would simply keep using the appropriate four-step methods to solve problems as they arose. Things were handled differently, though, once Dean took over leadership and put Mark in charge as director of quality and continuous improvement. Dean was the one who encouraged Mark to complete the TWI Institute forty-hour train-the-trainer programs for JI, JR, and JM so that he would have the knowledge to become the internal driving force for continuous improvement throughout the plant.

Even before Mark began teaching TWI classes, which is the purpose of attending the train-the-trainer program, he was getting great benefits from the knowledge and insights he gained there. "First," he explained,

> the timing was good because a major component of my new position is to focus on continual improvement. The TWI training provided me with the insight to understand the simplicity and effectiveness of the J programs at the time when I needed a good tool to not just drive improvement, but make improving a sustainable part of everyone's job. Taking the class gave me a greater understanding of the methodology and how the three modules of JI, JR, and JM work together to drive continuous improvement at the operator level.

Once Mark began teaching classes, he realized that his role as a TWI trainer meshed seamlessly with his responsibilities toward generating continuous improvement in the culture. "I soon realized," he said, "that although I learned how to use each four-step method in the ten-hour classes, it was not until I took the forty-hour trainer training that I began to understand how the nuances within the standardized delivery drove the learning process. The interesting thing about becoming a TWI trainer is that the more I deliver the training, the more I appreciate the importance of follow-up as we were taught in the forty-hour JI trainer program. That is when I began to understand there is a lot more to TWI than employee training."

"As an engineer I have always been drawn to simple effective things like cycling," Mark says as he relates TWI to his love of cycling. "What I love about cycling is the simplicity and efficiency of the bicycle. Using this as an analogy, I could just get on my bike and ride. But knowing the science behind this simple yet elegant machine, and the knowledge of how all these parts come together to perform a function, gives me an even greater appreciation of it. Most importantly, this understanding enables me to get the best performance from my bike. The TWI train-the-trainer training gave me the depth of knowledge of the thought process behind the methodology. With that understanding I got the 'why' and 'how' TWI should be used to get the best performance."

TWI has become part of the daily management system at Nixon Gear, and TWI is also written into their ISO procedures as a tool for preventive action and continual improvement. It is also an integral part of their continual process improvement strategy, as outlined in Figure 2.7.

Although the company has not yet created a program or system to reward or celebrate improvements, management does broadcast them at employee meetings and makes some "cheerful noise" to maintain the momentum, until they have an idea system in place in 2010. The plan is to continue promoting the usage of TWI by

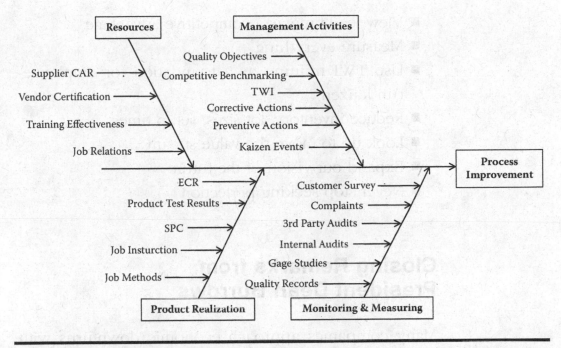

Figure 2.7 Continual Process Improvement Strategy. (Reprinted with permission of Nixon Gear, Inc., Syracuse, NY.)

demonstrating how everyone in the company benefits from its usage. And, as newly minted employee-owners, it is abundantly clear to everyone at Nixon Gear that each one of them has a stake in seeing to this success.

One final note: Mark's responsibilities were recently increased as manager of quality and continuous improvement at its Oliver Gear operation in Buffalo, New York. Beginning with ISO registration in 2010, the next phase of its continuous improvement plan is a full implementation of TWI/Lean at Oliver Gear.

The Nixon Gear Outline for Success

- Have a sense of urgency
- Plan carefully, execute swiftly
- Define and follow a process

- View lead time as a competitive advantage
- Measure everything
- Use TWI training as a foundation for employee-run kaizen
- Reduce inventory, lot sizes, setup times
- Look up to move the value stream
- Expand our vision of the future
- Never stop seeking perfection

Closing Remarks from President Dean Burrows

Many companies approach economic downturns with the focus to reduce costs, reduce investment, and "hunker-down" for the impending storm. At Nixon Gear we used this as an opportunity to invest in the future of our company, employees, and customers. We weathered this storm with smart investment, strong cash management, and self-funding the improvements. Through the acceleration of Lean and the resurging use of TWI, we exit the recession with a stronger balance sheet than when we entered it. Although our volumes were down 40%, our stock value dipped only 5%. The improvements that were driven during this downturn, through Lean and TWI, allowed us to double our cash on hand, reduce our inventories by 43%, reduce lead times by 50%, while improving our margins. As we exit this recession, we find ourselves better positioned to exceed our customers' expectations and to capitalize on business opportunities.

By dedicating the resources required, and committing the organization to implement the vision, success is inevitable. We may be busy, but we are never too busy to improve. We plan two to three kaizen events each month. As the improvements are implemented, they are locked-in and sustained with JI. In between our planned kaizen events, we coach employees on how to use JM to further drive the organization forward. Having a company resource and a well-trained workforce on TWI has made a measureable difference and distinction in our business. If you do not dedicate the time to improve, you will always find an excuse not to do it. You will be too busy, not have enough people, or create another excuse. As we improved our processes and freed up resources, thanks in good part to TWI, those resources were then used to improve other processes. It is a self-funding process.

Chapter 3

Getting TWI to Take Root in an Organization: TWI as an Integrated Process

The Nixon Gear story was presented first in this book of several case studies because it demonstrates a typical pattern of how companies come to TWI and struggle to get it to take root in the culture of the organization. Not all of the companies who see the need for TWI are ultimately successful in its implementation, as was Nixon Gear, and in spite of the great value of this program, it can go the same route as so many other good programs—another "flavor of the month." It shouldn't be this way, and indeed, that is the purpose of this book: to show how good companies have successfully integrated the basic skill sets of TWI into the daily functioning of their operations.

As we see it, there is no one correct implementation method or model, just as there is no one fixed culture or success formula for any company or business. Each business culture has its own unique history and characteristics that it brings with it when, for a variety of reasons, it decides to implement the TWI skills. Because the TWI programs are so fundamental, though, they can fit into any form or style of business, just as addition and subtraction fit into any mathematical formula, no matter how complex. Though these various formulas do different things in different ways for different purposes, they all still rely on basic mathematics to achieve their goals. And just as one organization may build aircraft carriers while another delivers healthcare, both systems still rely on the basic skills of instruction, leadership, and improvement to get their jobs done well on a consistent basis.

As seen in the Nixon Gear case, and as we shall see in the Donnelly Custom Manufacturing and Albany International cases, their efforts ebbed and flowed as they learned, step by step, what they needed to do in order to achieve, through TWI, the results their businesses demanded. It was not a simple, straightforward path, but one that evolved as they themselves and their organization grew with experience. They had to learn from their own unique circumstances the correct way of getting TWI to stick. At times, and this is a consistent comment we hear in most implementations, their practice of TWI actually helped to show them what needed to be done next. In other words, it was only after they began implementing the TWI disciplines that they saw where the program needed to go in order to deliver on its promise.

While each company's story is different, there are some consistent themes that we have found facilitate the process of getting TWI to take root and become part of the way the company operates—to become part of the company culture. The first lesson, and the most fundamental, is to take a leap of faith and trust in the basics.

Going Back to Basics

It seems odd to all those who understand the program, because they cover the most basic of skills needed to supervise in an organization, but it is oftentimes quite difficult to convince a company's management of the need for doing TWI. Just talking to them usually does not do the trick. It's like telling a bank manager that his staff needs to go back and learn their multiplication tables. "Of course they know how to do that," the manager will insist. "And besides, they have their calculators." Not only have we lost the art of basic training, but we remain consistently in denial that our people might not possess the skills necessary to do their jobs effectively.

However, what does work in getting management to recognize the need for basic skills is showing them examples of how poor training is leading to the unsatisfactory results they are currently getting. When we demonstrate the fire underwriters' knot, for example, just as it was done in the 1940s when the JI course was developed, they can see with their own eyes the results of just telling a person how to do a job or just showing a person how to do a job, typical ways people in their organization are "trained" to do their jobs today. It doesn't work well. When we show them the correct way of instruction using TWI Job Instruction, they then recognize that these basic skills are lacking.

Unfortunately, the two stories Patrick likes to tell about the need for basic skills come from two disgraced sports stars, now and then: Pete Rose and Tiger Woods. But because both of these athletes were the very best at their game, they provide good examples of how practicing the basics is what the very best do to stay in top form. It was said that, in his day, when Pete Rose was not happy with his hitting in a home game, he would have the stadium lights turned back on in Cincinnati and would do more batting practice—after everyone else had gone home. And when Tiger Woods was in a slump several years ago, he got a coach and went to work on his swing—just like we do when we want to improve our golf game—and he began his long march of tournament victories. So there is no shame in going back and practicing the basics. That's what the very best in the game do.

And that's what TWI is all about—the basics. To get TWI to take hold, then, we first must have faith in these basics and know that there is wisdom in the ages. The TWI dialogue, written in the 1940s, states that "there is nothing new about these methods. Good supervisors have been using them all along." When management understands and supports the need for basic skills development, they can begin to create a strong foundation for the other Lean processes and tools to begin working and providing results. We'll explain how this works in the next chapter.

While it is usually difficult for management to understand the need for these basic skills, just the opposite is true with front-line supervisors. A very common comment we get in our TWI classes is, "Why didn't you teach me this when I started as a supervisor? It would have made things a whole lot easier." This response indicates that supervisors recognize the need for these kinds of skills, and when the proper training is not provided, they have to learn them the hard way—in the school of hard knocks. Even experienced supervisors, who make up the bulk of our classes, consistently say that they can continue to learn and improve on these necessary skills.

To use another sports analogy, when Patrick was doing TWI training in Alabama in November 2008, there was a lot of talk about football as the Alabama–Auburn game was coming up that weekend. Taking a cue, he explained how TWI is just like blocking and tackling: "Even when you have the most talented players on the field, if the team can't block and tackle, you lose the game." Everyone in Alabama, where football is king, understood what that meant, and the members of the class nodded vigorously in agreement.

The Three-Legged Stool

At the end of *The TWI Workbook* we explained very briefly how the three parts of TWI—Job Instruction, Job Relations, and Job Methods Improvement—are just like the legs of a three-legged stool: if you take one away, the stool falls over. In other words, the three programs work together and support each other when TWI is introduced as an integrated program. Since we wrote that simple metaphor, many companies have gone on to implement two or three of the TWI programs and are proving its validity. We are now in a better position to show how this synergy works and why TWI as an integrated program builds a stronger foundation when it becomes part of a company's culture.

Job Instruction as a Stand-Alone Program

Before we look at the strong relationships between the three TWI programs, though, we have to recognize that many companies, some of the best in their fields and a few of whom are represented as case studies in this book, elect only to implement the Job Instruction portion of TWI. Each of the TWI modules can stand alone as a practical and useful skill, and you don't need the other two to make any one work. In particular, Job Instruction was the first module to be developed, and it is the foundation upon which the other modules are based on. Not unexpectedly, it has proven over time to be the most popular of the three.

Job Instruction is fundamental to the success of any organization, and its results go directly to the bottom line. Having a well-trained workforce that understands their jobs and can perform them with accuracy, precision, and timeliness is what keeps a company strong and profitable. What is more, as organizations of all stripes energetically study Lean methodology and begin to recognize the need for enforcing standard work, they come to TWI JI as the means of achieving and sustaining that standard performance. As Art Smalley, one of the first foreign nationals to work for Toyota in Japan who is a Toyota Production System (TPS) expert and JI trainer, points out:

> Basic stability starts with a well trained workforce. Fortunately employees tend to know their jobs very well or we would all be in serious trouble. However, Toyota in the 1950s learned some basic techniques about supervision in production and how to further

improve the skills and capabilities of work teams. Specifically, they adopted an industrial training program that the U.S. used during WWII called Training Within Industry (TWI).*

A variety of works that have come out in recent years on The Toyota Way have emphasized and described Toyota's continued use of the TWI methods, in particular Job Instruction. It is not surprising, then, with the high level of interest in emulating Toyota's manufacturing methods, that Job Instruction has become the focus of much interest. The intent here, though, goes much deeper than just imitating Toyota. There is a reason why JI was so influential to the Toyota Production System and why companies trying to emulate Toyota see the need for good instruction skill. The reason is the necessity for standard work as bedrock to all the other Lean tools and practices.

Companies have been attempting for years, like we saw in the early stages of the Nixon Gear story, to implement Lean only to find out that they missed the most fundamental piece: achieving and maintaining standard work. They profess to understanding standard work, but at the same time, they also admit that in reality the jobs being done at their plants are performed differently, "the way the operators like to do them." In almost every company, different operators doing the same jobs do them differently, and even when done by the same operator, the method changes and varies from cycle to cycle.

We will discuss how JI provides the tool to get to standard work in the next chapter, but the point here

---

* Art Smalley, Basic Stability Is Basic to Lean Manufacturing Success, Art of Lean, Inc., www.artoflean.com.

is that the overwhelming need for Job Instruction means that many companies striving to make Lean a reality and not just a slogan put a singular and concerted focus on implementing just the JI program alone. We don't think that's a bad approach at all, and when done well, these companies achieve spectacular results, as we saw in the ESCO Turbine Technologies–Syracuse case, and as we shall see in more case studies to come in this book.

There is one big caveat, though, to these successes. In practically every instance where we saw JI implemented successfully as a stand-alone program, companies already had a strong culture of human development in place. These companies valued their employees, listened to their opinions, and put emphasis on strong human relations in the work environment. In almost all cases, as well, they had an improvement program under way that emphasized and rewarded the participation of all employees, especially the operators. Therefore, people in these companies already had as strong sense of involvement in the daily operations even before JI was implemented. The Job Instruction skills complemented the strong and steady leadership of the company's management.

When this is not the case, however, we find that what is needed, even more fundamentally than standard work, is good leadership. TWI, when used as a fully integrated program, provides a solid foundation for leadership where employees can look to their supervisors with respect, knowing that they possess the skills to provide them with what they need to succeed in their work. If they do not trust their leaders enough to follow their instructions, then the best instruction method in the world will not get them to adhere to the standard methods. If they are not *following* these instructions, then their supervisors are not *leading*.

The Relationship between JI and JR

In our JI classes we repeat again and again the core concept of the method: *If the worker hasn't learned, the instructor hasn't taught.* (Many people coming out of Toyota have said they always thought this line was developed by the Japanese and are surprised to hear that it was actually Americans who coined the phrase.) Inevitably, though, in many classes there will be one supervisor who will protest and say something like: "Just wait a minute here. You can't put that on me. You can't tell me it's my fault if they don't learn. If you only knew the kind of people I have to supervise...." When we challenge them on this, they almost always fall back to the position: "Even if I had a good instruction method, they still wouldn't listen to me."

That is correct. But this isn't a problem of instruction, it is a problem of leadership. As we intimated just above, a leader is a person who has followers. If people are not following your instructions, then you are not leading them. Therefore, leadership is a foundation to good instruction. The stated goal of the Job Relations component of TWI is to gain the dedication and cooperation of people in getting the work done. When we apply the JR method of developing and maintaining good relations with people, only then, when people *want* to do their jobs correctly, can we expect good results from our instruction effort. If people really do want to perform their jobs properly and they still can't, then indeed we can say, "If the worker hasn't learned, the instructor hasn't taught." Here we can clearly see the relationship between JI and JR in that the leadership element of JR, the skill in leading, supports the effective use of the skill in instructing.

You may wonder what this skill in leading looks like in actual application. Here is a story we hear repeated continuously. How many times in the course of an ordinary week, or even a day, does a supervisor find himself or herself in a heated argument over a serious personnel situation, a people problem that entails strong emotions like anger, fear, disappointment, vengefulness, etc.? Typically these serious cases end with someone in higher authority splitting the difference and trying to come up with some type of fair solution so everyone can stop bickering and get back to work. But the result is that each person involved goes away unhappy and disgruntled, and the original problem, still left unsolved, continues to fester and grow only to come back later in a larger and more virulent form.

With supervisors now trained in JR, in the midst of all the shouting and finger pointing, one of the participants in the discussion will pull out their yellow JR card and ask the group, "So, what is the objective here?" This is the first question of the JR method to solving people problems. Once everyone gets focused on the method, immediately all of the emotion is sucked out of the room and people calm down and revert to a more effective way of problem solving.

After agreeing on what they want to see happen, they begin assembling the facts, Step 1 of JR, and in particular try to capture the opinions and feelings of the individuals involved in the problem. Based on these facts, they then move on to considering what possible actions might be taken and evaluate these various options. That's Step 2. Once they pick the actions they think will bring the problem to a successful resolution, they can quickly and easily move on to Steps 3 and 4, taking the action and following up to check if good results were obtained.

This is just what happened at Nixon Gear when Mark Bechteler was facilitating a team of several people, four of whom were highly skilled CNC (computer numerical control) machine operators. One of these operators brought in a JI breakdown on how to balance a large part when loading it on a grinding machine. As it turned out, all four operators loaded that same machine differently, but all produced good parts in the required time. Discussion on which operator did the job "correctly" became heated and extended into the next team meeting. One of the operators came to the third meeting fed up with the impasse. He pulled out his yellow JR pocket card, set it on the table, and asked, "What is our objective?" The team settled the issue within thirty minutes, and the JR method has since been used to reach consensus at all team meetings.

This scenario gets repeated over and over again at companies that embrace the fundamentals of good job relations, and the outcome is clear: the organization can get to a good solution where everyone involved is satisfied with both the process and the actions taken. When people can walk away from an emotionally charged situation and feel good about what happened, this is the first step toward building strong and trustworthy relationships. When the employees directly involved in these situations have their opinions heard and their feelings considered, regardless of the outcome, they feel they have been treated more fairly and are much more likely to respect the leadership that brought them to the final conclusion of the problem.

This is how good relations are built, and the ensuing trust contributes directly to the successful implementation of standard work. After all, why would someone follow a system if he or she didn't have respect for and trust in that system? A worker's direct supervisor is the living embodiment of that system, that standard of work, and if employees don't trust their supervisor to do right by them, why would they follow those instructions? Instead, they will think they have a better way, believing that the supervisor's method, the "standard," is not made with their best interests in mind. If you want people to follow standard work, first you have to build their trust in you as their leader. As we said earlier, if they are not following, then you are not leading.

The Relationship between JI and JM

The connection between Job Instruction and Job Methods Improvement is pretty straightforward and clear: after a new method is developed in JM, use JI to teach it to the operators and ensure that it gets used correctly and consistently. The benefits of a new method will not be realized until the people doing the work put it into use. And if they don't learn these new methods well, they will not be able to perform them as developed.

Many, if not most, companies that embark on a Lean journey find that, after some initial gains, they stagnate or fall back. The sheer impetus of the kickoff campaign for change leads to early successes as "low-hanging fruit" opportunities are found and improvements made. But before long, operators fall back on the old methods that are more familiar to them and, even before the posters begin to fade, continuous improvement proves not to be continuous. So while their efforts are put in the right places, the ability to maintain the higher standards is missing and enthusiasm for the effort wanes.

Numerous companies come to TWI Job Instruction because they are seeking to sustain the gains they have made from their Lean programs—and the JM method of finding improvement fits right in with these Lean activities. In the same way, though, that operators revert to pre-Lean procedures and habits, if they are not taught well the new methods found using JM, these improvements will also not be maintained and, in the end, benefits will not be realized. Keep in mind that good job instruction technique is not merely explaining the new method to operators, but also getting them interested in and motivated to do the work—that is covered in Step 1—and then following up with them to ensure they are using the new procedure correctly and consistently—that is Step 4. Furthermore, when reviewing the new method being instructed in Steps 2 and 3, by giving the reasons for the key points to the procedure, the learner is able to understand the benefits to the changes being made and can more easily accept and adopt the new method. When people understand why they have to do things a certain way, they will be more conscientious of what they are doing and remember to do it that way all the time.

In this way, JI is the tool by which improvements made in JM are put into use. Without it, improvement activity becomes a futile exercise in which good ideas are squandered and left to go to waste.

On the other hand, JI is also a necessary prerequisite to the improvement activities of JM. In most improvement schemes, we begin by analyzing the current way the job is being performed. Likewise in JM, we first get the details of the current method so that they can be questioned to find ideas for improvement. However, if there is variation in the way the job is being performed, we will never be able to perform a consistent analysis of this current method. In other words, each time we look at it to capture the details, it will be a different series of details. Consequently, we will not have a solid baseline upon which to get ideas and build our improvements.

When we first began reintroducing TWI back into the United States in the early 2000s, we were working with a large auto parts company that insisted on starting their TWI training with JM. The class participants were giving their demonstrations, in which they practiced using the JM method on jobs in the plant. As one of them described the current method details of the job he had studied, a few Industrial Engineering staff who were participating in the training protested that the operators were not supposed to be doing the job that way. "But that's what I observed!" the presenter retorted. In the end, it turned out that the ideas he got by doing the JM exercise were things that had already been in the standard procedure—the procedure that was not being followed by the operators. The "improvement" was a rehash of earlier ideas that had not being implemented correctly or consistently.

Even in companies that declare they are enforcing standard work, and present many thick binders of standard operating procedures (SOPs) to prove it, when we go out and talk to people on the floor, almost invariably we find that there is great disparity in the way the work is actually being performed. Just because it is written down doesn't make it happen that way in the actual work. There needs to be a medium that bridges the gap between the ideal version of the standard, which is documented in the SOPs, and the actual application of that standard, which is the actions and behavior of the people performing the jobs. That medium, of course, is good instruction as carried out in the JI method. Without the stable performance we get from a well-trained workforce, improvement activities will not bear fruit. This discussion will be more fully developed in Chapter 4.

The Relationship between JR and JM

In the Job Relations module we point out foundations for good relations that, when applied on a regular basis, help a supervisor build strong relations with his or her people in order to prevent problems from arising. Perhaps the most challenging of these four foundations is to "make best use of each person's ability" because we have to find ways to let people grow by bringing out latent abilities even they may not be aware of. When this is done successfully, people get a sense of their own potential, gain more satisfaction and pride from their work, and view their own organizations as good places for them to work because they will continue gaining greater fulfillment in their work and advancement in life. You will see a good example of this in Chapter 6, the Albany International case study, where Jennifer Pickert became a dedicated JI trainer and Jamie Smith became a certified JI Instructor teaching the ten-hour class after each was "discovered" while participating in the TWI ten-hour classes and activities.

Lean initiatives seek to tap into this potential by allowing employees to participate in the improvement and design of their own jobs. After all, as these programs promote, who better to understand the current status and needs of a job than the person doing that job day in and day out? The JM program recognized this effect back in the 1940s when it was developed. The method has always been taught to front-line supervisors who are encouraged to get their people involved in the process. By "working out your ideas with others," the method prescribes, using an operator's ideas is a way of getting a "satisfied employee," which is "perhaps more important than the idea itself."

In this way, JM provides the means by which we can continue building stronger relationships with our people, which is the stated objective of the JR program. Once this is done, the other three foundations for good job relations can easily be applied. Successfully creating and implementing improvements provides an opportunity to "give credit when due" to those people who participated, and it makes it easier to "tell people in advance about changes that will affect them" because they have already taken a part in developing what those coming changes will be. Moreover, because the operators have had a hand in deciding the new method, it will be that much easier for them to learn and put them into use because they have taken ownership of the method by becoming part of the team that authored the new standard. Finally, we can "let each worker know how he or she is doing" based on a standard he or she helped to create.

On the other hand, in much the same way that JR helps to build motivation for following instructions given in JI, it will also help to build morale around participating in improvement activities such as JM. It is an axiom of good Lean initiatives that overall success is dependent on everyone's involvement and contribution. But for a variety of reasons, many operators are skeptical of and resistant to making a whole-hearted commitment to these efforts. They may be fearful of the results, worried that their ideas will lead to their own layoff or dismissal. Or they may simply be tired of yet one more program that gets a few weeks, or months, of attention only to be discarded and forgotten when the promotion dies down.

Using the JR four-step method of handling problems, supervisors can get the facts and find out why certain individuals are not willing to participate and give ideas for improvement. By talking with them and getting their opinions and feelings, supervisors may find that there are deeper reasons to their ambivalence toward job improvement other than our usual assumptions about "bad attitude" and "lazy workers." For example, they may find that the person did give an idea in the past for which he or she never received even a thank you, and so decided "to never make that mistake again." Moving on to Step 2, weigh and decide, supervisors can take these facts and come up with a variety of possible actions to correct the problem, such as offering that person a position to lead an improvement team and take responsibility for making sure *everyone* on the team gets credit for their work. Taking action and checking results rounds out the

Table 3.1　How the TWI Programs Work Together

	JI–JR	JI–JM	JR–JM
Relationship	Leadership skill (JR) is foundational to instruction skill (JI).	Good instruction (JI) both precedes and follows improvement efforts (JM).	Improvement activities (JM) enhance strong relations (JR) and vice versa.
Principle	People will not follow our instructions if we do not lead them well.	Work processes must be stabilized before they can be improved.	Direct involvement in designing jobs inspires positive work ethic.
Benefits	Standard work is adhered to when people want to follow good instruction.	Improved methods will stick when they are taught properly.	Kaizen results multiply exponentially when people are actively involved.

method, and results can be measured by seeing whether the attitude of the individual has changed, whether his or her participation has made an effect on the output, and whether others in the group have responded positively to the person's change in behavior.

Table 3.1 summarizes the way in which each of the TWI programs supports and enhances the others.

What Is the Correct Order to Roll Out the Three Programs?

Once companies see the value of implementing two or all three of the TWI methodologies, the next logical question is: In which order should they be rolled out? As stated earlier, each of the three programs can stand alone, so no one implementation is required before you can start another. You can begin with any of the three modules and, depending on company needs, a good argument can be made for each one being the logical place to start.

When we first began introducing TWI back into the United States in early 2001, many companies, both large and small, were extremely enamored with Job Methods Improvement, eyeing the possibility of gaining immediate benefits in productivity and profitability. Though they could see the value of good job instruction, they felt the first need to address was ridding themselves of inefficiencies and waste, which the JM program went after quickly and effectively. Moreover, they felt that there was no need for the Job Relations course because they had already taken their supervisory staff through countless seminars and training courses on how to deal with people and felt they had these skills covered. Job Methods was the TWI course they wanted first, and perhaps they hoped that it was all they would ever need.

More than just that, though, many of the people we were working with at the time were actively engaged in the early stages of Lean implementation and were focused extensively on kaizen and other improvement activities. For them, fixing the workplace was the initial order of business, and besides, they said, "What is the sense of teaching an inefficient method?" First, they insisted, before teaching the job, the most effective way of doing that job had to be found. Otherwise, they would find themselves teaching less than ideal job standards.

In spite of our arguments for doing JI first, these debates continued for a few years until books like *The Toyota Way Fieldbook* and *Toyota Talent* came out and described the need for standard work as a foundation to continuous improvement. This had always been a part of the Lean ideology, but it was not clearly understood. In particular, the way to get to standard work, using TWI's JI component, was clearly introduced in these Toyota texts, and people began to understand, perhaps for the first time, that a lot more work was involved here than simply composing and issuing standard operating procedures. As it turns out, Job Methods should be the *last* program introduced in a TWI rollout, and as several of the case studies in this book will illustrate, that is what has proven most effective.

JI before JM

We explained above, in the section on the relationship between JI and JM, how important it is to have stable work performance in order to be able to analyze a job for improvement. This is perhaps the strongest reason for doing JI before JM, but there are others, and they are very compelling. For one, it is fairly common for supervisors to *not* know or even be familiar with the jobs they are supervising. The question quite frequently arises when doing a JI class as to how a supervisor is supposed to break down and teach a job he or she does not even know.

Rod Gordon was the vice president of manufacturing at McCormick and Company, Inc. when he found TWI on the Internet late one night while struggling to come up with solutions to sustainability in the Lean implementation he was tasked to direct. Rod had recently come in from Pepsi, where he was a hardened veteran of the cola wars and was determined to change the culture at McCormick to respond to the pressures of a modern global market. Though McCormick is the dominant producer of packaged spices for both the industrial and consumer markets, it was just a matter of time, they knew, before other countries, such as India or Viet Nam—and other places where the actual raw materials are grown— found ways of grinding up spices and putting them into small packages. They had to get lean if they were to maintain their leadership in the industry.

After our initial pilot training, Rod was sitting quietly in the wrap-up meeting, where managers and other leaders were critiquing the course just completed, when one person asked, "Does this JI mean that supervisors have to know the jobs they supervise?" Rod suddenly perked up, slammed his fist on the table, and declared, "Damn right, they do!" He then went on to explain how, as a young Yale graduate on his first job at Proctor & Gamble, he had to demonstrate that he could perform every job on a line before he was allowed to be the supervisor of that line. That's how good companies do it, he explained, and if the supervisors at McCormick didn't know how to do the jobs they were in charge of, then this would be a good opportunity for them to learn.

The big lesson here is that if we expect our front-line supervisors to lead the charge toward becoming Lean, if we expect them to find and take out waste and to discover improvements to get better productivity, if we expect them to redesign processes to create better flow, if we expect them to do any of these Lean processes, they simply must have a good understanding of the work being performed. How can we expect someone to find waste or improvement in a job they don't even understand? It is certainly true that they will never know a job or be able to perform it as well as the person who does it every day, and that they will need that operator's support, insight, and guidance when improving the job. But to lead this endeavor, the supervisor must have a fundamental knowledge and respect for the work in order to point the improvement effort in the right direction and ensure that the Lean tools are properly applied.

The supervisor does not need to be the best performer of a job in order to teach it, just as the coach of a sports team does not need to be the best player on the field in order to give instructions to the team members. No one expects the coach to go out and take over for a player who is performing poorly. But the coach does need to understand how the game is played and to instruct players, even the superstars, on how to perform their roles in order to win the game. That includes going back and relearning the fundamentals when needed.

Job Instruction, then, serves to both build the skills of the operators in doing the jobs and create expertise in the supervisors in directing the work. Efforts from both these angles build a strong foundation upon which continuous improvement activities will grow and flourish.

One more thing. There is an improvement element to JI that gets an organization moving down the kaizen path even as it is working on enforcing standard work through good job instruction. Because there is variation and instability in the way work is done, teaching every-one to do the job the "one best way" means that all of the "not so best ways" will be taken out. The truth of the matter is that each worker has his or her own spe-cial skills or knack for doing the jobs he or she does. These are learned, oftentimes unconsciously, over time and through long experience. When supervisors quiz the various operators of a certain job on how they do it, they can then consolidate all of the best techniques into one standard that everyone can use. Moreover, by including some of the techniques each worker uses in the final method, it will be easier to get buy-in from all of them to follow the standard because each has made a contribution.

For all these reasons, it is extremely effective to do a full JI implementation before moving on to JM. Many companies, in fact, will go several years reaping the ben-efits of Job Instruction and continue no further than that. When we ask them why they don't move on to JM, they claim that the rewards of JI are so great they want to keep them coming in without disruption and will get to JM when they are ready. That was the pattern at ESCO Turbine Technologies–Syracuse that credited the JI pro-gram for a 96% reduction in defects in their wax depart-ment from 2002 through 2004 while reducing training time in that department from two months to two weeks.[*] The plant then found the need to introduce Lean manu-facturing in 2005 and followed that by introducing Job Methods as a tool for continuous improvement.

---

[*] Patrick Graupp and Robert J. Wrona, *The TWI Workbook: Essential Skills for Supervisors,* Productivity Press, New York, 2006, case study on CD.

JR before JI (or the Other Way Around)

The next question, then, would be whether to implement Job Instruction before or after Job Relations. As we explained above in the section on the relationship between JI and JR, leadership is a foundation of good job instruction. So logically we should teach JR before JI. If strong relationships built on trust and respect are not firmly established up front, instructions will not be followed with precision or enthusiasm, and the result will be all the problems we find when people lack training or are not trained well. So the success of a JI implementation depends on whether the leadership component is already in place.

The reality of the matter, though, is that the vast majority of companies we work with begin with Job Instruction. Though they may understand the compelling need to do JR up front, because JI provides quick results for clearly defined problems, JI is relatively easier to launch with strong top management support. In other words, it's easier to "sell" the organization on the need for JI than for JR. As with all initiatives of this scale, management support is a vital ingredient to a successful TWI rollout and must not be overlooked—it is an element worth making sacrifices for in order to obtain. But as we noted, it is very often difficult to get management to see the need for JR since they have already spent a lot of money and time trying to educate their supervisors on good people skills.

In any case, JI is not a bad place to start since training is always a continuing and compelling need for any organization. Because of that, when JI is introduced in a proper manner, good results come in quite quickly and, when rolled out well, momentum begins to build for using JI as a standard way for training in the company. Moreover, JI contains a very strong dose of human leadership components (see Chapter 7) that cover many aspects of good job relations, and so that part of the equation is not neglected. In a company that does a good job of emphasizing and promoting strong job relations, then, JI proves to be an excellent starting point.

Having said that, our experience has shown that the upper management of many companies gives themselves more credit for creating strong human relationships in their organizations than they may actually deserve. This is not to say that they do not want or strive for a strong people-oriented culture. They believe in what they preach and assume that people are being treated correctly at all levels of their organizations. But that may not necessarily be the case, and it is easy for management to insulate themselves from the true reality of what is happening in their own companies. People in their organizations may never inform them of the actual troubles they are experiencing on a regular basis because they feel this kind of "bellyaching" about seemingly petty people problems will not be seen in a good light by their superiors. When these minor problems eventually grow into major ones, which they almost always do, then management comes in with the fix because that is what they do: they solve problems. But they never get to the underlying issues that led to the problems, which continue to fester.

Front-line supervisors, on the other hand, are inherently aware of their own lack of skill in dealing with people issues because they have to face them every day. Right from the start, as we reported in *The TWI Workbook*, supervisors could see the need for JR, even more than JI. In spite of all the training they may have had previously on communication techniques and understanding human personalities, they still do not have the skill to resolve people problems in a way that builds strong relationships and good morale in the workforce. In effect, they are drowning in people problems, which prevents them from achieving their goals for producing good quality products and services.

What can be done about this mismatch in perceptions? Instead of trying to convince top management that they really do need JR, what usually happens is that companies go ahead and start with JI and make a concerted effort to make JI the standard for training in the company. After that, once the organization begins to see the power of good training, it also begins to see the need for JR. In other words, once the environment begins to change from one of blame—"the operator failed to do the job correctly"—to one of responsibility—"if the worker hasn't learned, the instructor hasn't taught"—then the burden falls on the supervisor to ensure he or she has motivated the operator to do a good job. This is just the goal of JR, and it becomes very apparent in this new paradigm of supervision that this skill is lacking.

As good instruction skills begin to be implemented and the need for JR becomes apparent, many companies doing JI move quickly to implementing the JR component. With the strong feedback and enthusiasm management sees for the JI implementation, not to mention the good results they begin receiving, it is much easier to acquiesce to doing JR because that is what supervisors are now asking for. JI and JR work closely together and go hand in hand in providing a strong leadership program that develops skilled people who are dedicated to doing a good job each time it is performed. Whether JI comes first or JR comes first, these two programs should be done closely together because of this tight relationship. Table 3.2 sums up these points on the ideal sequencing and timing for a strong TWI rollout.

Table 3.2   Ideal TWI Deployment Sequence

1st—JR	By first building strong relations, supervisors gain trust and respect from their people, who will then want to follow their good instructions and participate in improvement activities.
2nd—JI	Good job instruction will then create stability in work processes, which will become the foundation for developing true standardized work. Only then can processes be analyzed properly to find improvement.
3rd—JM	Once good morale is established around following standard work processes, efforts at getting everyone involved in improvement activities will flourish. Good instruction practice will ensure improved methods are taught well to employees who want to follow the new standards.
JR–JI switch	Job Instruction can be introduced first if strong human relations exist. Even in that case, the need for Job Relations becomes quickly apparent, and JR should be implemented soon thereafter.
Timing	Job Instruction and Job Relations should be rolled out fairly simultaneously, with just enough time in between to master one before moving on to the next. Job Methods can be implemented much later (nine to twelve months or longer) when processes stabilize.

Conclusion

Although every company may take a different approach to rolling out TWI, these fundamental skills apply to any company's work in any industry. Because they are so basic, they are oftentimes overlooked. When Rod Gordon of McCormick was first discussing with us his plans for TWI within their Lean strategy, he exclaimed somewhat incredulously, "How did we miss this part?" Most executives are not that honest with themselves, but the fact is that we become so enamored, or overwhelmed, with the difficult things that need to be done, we leave behind the basic things. Getting back to basics is always a good place to go when the going gets rough and this essential insight is needed to get TWI to take root in an organization.

While the TWI programs are effective as stand-alone programs, especially JI, they become an integral part of the culture of a company when they are used together. Each program complements and reinforces the other. They form a core set of skills that, when used in tandem, make leadership a reality and not just a philosophy. Knowing the relationships between the programs and an effective sequence to rolling them out will contribute to getting these skills to stick.

The final question that remains, once we decide on a sequence of which programs we want to implement, is the strategy for rolling out that sequence so that it embeds in with the organization and becomes part of the daily culture. These are more implementation tactics than specifics of the TWI programs themselves, and they have to fit into the larger context of what the organization is doing, and trying to do, with its overall Lean implementation. So in the next part of this book, we will look in more detail at how TWI relates to the core Lean values and why TWI provides the foundation for success in Lean. Several case studies will show how companies are doing that today. Then, in Section IV of the book, we will go into greater detail on how to manage a successful rollout of the TWI methods.

II

TWI'S CONNECTION
TO LEAN

TWI as an Integral Part of Strategic Lean: Standard Work, Continuous Improvement, Respect for People

Taken collectively the 3J courses can provide a very important base for a work team in a lean program that will sustain. At the heart of every system are people and their collective beliefs and behaviors. The TWI material played a powerful shaping force early on in Toyota's improvement journey. Instead of using value stream maps, kaizen events, or pull systems which all came later in Toyota's journey these basic courses were foundational elements in making the Toyota shop floor a more stable and predictable environment. I urge all parties interested in these materials to make them an integral part of your improvement journey.

—**Art Smalley**[*]

---

[*] Art Smalley, Basic Stability Summary Document, Art of Lean, artoflean.com, 2008.

Through the work of Toyota veterans and scholars like John Shook, Jeffrey Liker, Art Smalley, and David Meier, it has become abundantly clear that, at Toyota, before there were value stream maps, *poka yoke*, or Just-in-Time, there was TWI. These TWI foundations helped form the core elements of what we know today as the Toyota Production System (TPS), but most importantly, what we need to remember is that at Toyota these fundamental skills didn't go away as they progressed to more sophisticated techniques. In fact, these fundamental skills are what make the higher-level tools work, just like you need addition and subtraction to do more advanced mathematics, and just like you need to first learn to block and tackle to win the football game, no matter how talented your quarterback may be. As we discussed in the last chapter, practicing basic skills is what the best practitioners in any field of endeavor do to stay on top.

There is a Japanese saying that "even a monkey sometimes falls out of the tree," which implies that even an expert can make a mistake. As we write this chapter, Toyota is perhaps facing the most difficult challenge of its storied existence as it halts production of key models and recalls millions of vehicles across the globe to repair faulty gas pedals. How could the world's best auto company with a bulletproof reputation for quality be faced with such a colossal failure on the broadest scale imaginable? That story will be written by others, but we believe that the fundamental principles of good manufacturing that Toyota has developed and practiced will continue to be the gold standard for production excellence, and that the company will regain its premier position by going back to what it has practiced all along—standard work, continuous improvement, and respect for people.

Nevertheless, the development of Lean production has come a long way since *The Machine That Changed the World** introduced the world to Lean and to the practices of good manufacturers like Toyota. Certainly Toyota has exemplified that standard of excellence, but with companies all across the globe implementing Lean tools and concepts, the practice of Lean has spread well beyond only what is done at Toyota. And, as Akio Toyoda, president of Toyota, said in an interview with Japan's daily financial newspaper, the *Nihon Keizai Shinbun*, "When our annual output surpassed 6 million units (around fiscal 2002), we started moving so fast that we didn't have time to bring up our workers the right way. Toyota's training of workers to maintain quality control failed to keep up with the company's rapid growth."† As happens so many times in all of our lives, it seems that Toyota too must go back to the basics.

In this chapter, we would like to point out and explain how TWI supports the Lean fundamentals and why, without TWI, application of these tools so oftentimes turns into an exercise in futility. In particular, the core Lean concepts of *standard work, continuous improvement*, and *respect for people*, we feel, come right out of the TWI basics as they were developed in the 1940s. Moreover, we will show how applying TWI to an already

---

* James P. Womack, Daniel T. Jones, and Daniel Roos, *The Machine That Changed the World: The Story of Lean Production*, First Harper Perennial, New York, 1991.

† Toyota: Worker Training Lapsed Amid Fast Growth, SME Daily Executive Briefing, March 19, 2010, quoting an interview by Akio Toyoda in *Nikkei Shinbun*, March 7, 2010.

existing Lean initiative will supercharge that effort and help gain the return on that investment.

Achieving Process Stability

As we explained in the last chapter, it is a common misunderstanding that the Lean journey begins with improvement. "After all," new Lean initiates will insist, "the whole idea of Lean is to get better by eliminating waste, and we do that by making improvements to our current processes, right?" However, as Art Smalley teaches in his classes on basic stability, "I recommend that it is often better to put aside the 'sexy' items in lean such as *kanban*, standardized work, or value stream maps, and instead work on the more fundamental items" of creating basic stability.[*]

Lean Implementation Sequence

Toyota has been reluctant to publish or even endorse what it considers to be the right way to implement Lean because Toyota executives view TPS/Lean as a system of thinking that practitioners learn best by doing. When pressed, veterans of Toyota will suggest that there are preconditions needed for a Lean implementation to proceed smoothly: "These include relatively few problems in equipment uptime, available materials with few defects, and strong supervision at the production line level." Smalley also learned from interviews and conversations

---

[*] Art Smalley, Basic Stability Summary Document, Art of Lean, 2008, artoflean.com, p. 1.

with retired Toyota executives about stability how Toyota learned the hard way that in the beginning of a transformation a company needs stable processes before being able to succeed with the more sophisticated elements of Lean. What Toyota learned was that if we attempt to improve an unstable process, we will not be able to conduct an accurate analysis of the "current state" because it changes each time we observe it or when different operators do the same job differently. Even if we do find improvements under these conditions, our "new method" will just be one of any number of ways that the job will be done and the benefits of the improvement will be lost.[*]

Toyota uses the basic building blocks of manufacturing—the 4Ms of manpower, machines, materials, and methods—to establish a consistent and predictable process before going too far with the latter elements of flow and takt time that contribute to standard work. The key to achieving process stability, as passed on by former Toyota executives to Art Smalley, who in turn documented the process, was to concentrate on these four key elements:

1. Manpower: Basic stability starts with a well-trained workforce.
2. Machines: You must know your customer demand, the capacity of your process, and the actual average output.
3. Materials: Apply the tools of Lean to reduce waste to shorten the timeline from when the order is received until it is produced.

---

[*] Art Smalley, Basic Stability is Basic to Lean Manufacturing Success, www.lean.org, p. 2.

4. Methods: Have standard methods in place as a measure for how you are doing and to measure the impact of change.[*]

Everyone practicing Lean understands that standardized work is the bedrock foundation of good Lean practice. When companies study Lean, however, they jump right to improving methods and what they feel is creating standard work. For them, this means writing it down in detailed descriptions called standard operating procedures (SOPs) or standard work documents or work instructions. The idea is that these documents are to be read by the operators doing the work and followed so that everyone does it "the standard way." As we shall see in several case studies in this book, though, just because you write it down doesn't make it so.

What Smalley is saying is that before we can come up with a standard for the work that we can lock in and declare "the official way the work is done each and every time," we need to create stability around what is being done today. In other words, we need to use the JI breakdown method to determine the current best way we know how to do the job, train every person in that method, and most critically, develop a follow-up system so that we can make sure everyone follows that method, just as was done by ESCO Turbine Technologies–Syracuse (see Chapter 1). Once we have everyone trained in that method, and motivated to following these instructions, then we have a baseline to analyze that stable performance, to look for improvement opportunities,

---

[*] Ibid.

and if the method proves sustainable, to document it as the standard.

The three TWI skills at work in this process drive us toward the ultimate goal of standardized work. Job Relations creates the needed cooperation from the people who are to do the work so they are willing and motivated to follow the instructions they are given. Job Instruction continues with this effort and directs it to a specific job that is to be taught. First, though, the job is broken down and "the one best way" we know how to do the job is determined so that it can be learned quickly and performed correctly, safely, and conscientiously. Finally, opportunities for improvement can be mined on a daily basis by using Job Methods to improve upon the current best way that will ultimately result in new methods taught back to the operators using Job Instruction. Once the process is stabilized and we have consistent performance of the job across operators and shifts, we can proceed with the other three Ms of machines, material, and methods, as required, to write our standard work document.

Standard Work and Continuous Improvement

When going down the path to standard work, it is wise to keep in mind the advice of Masaaki Imai, who reminds us that the journey does not end when the initial work is done:

> In reality, there can be no such thing as a static constant. All systems are destined to deteriorate once they have been established. One of the famous

Parkinson's Laws is that an organization, once it has built an edifice, begins to decline. In other words, there must be a continuing effort for improvement to even maintain the status quo.[*]

Improvement activities such as value stream mapping and kaizen events that are focused more on process improvements may result in creating a better method. Of course, that is what kaizen is all about, but unless that method standardizes work, there is no way to be sure that you have in fact improved the operation, or if you just changed the way a particular person was doing the job at that time. Then, even when we perform organized improvement events, the improvements made begin to decline unless there is a continuing effort to maintain the new levels of performance. This is shown on the sawtooth improvement curve in Figure 4.1.

This brings us full circle, then, to our original chicken-and-egg dilemma: Which comes first—standard work or improvement? It should be clear by now that without a stable process, one in which all operators are doing a job the same way across all work shifts, it will be difficult to analyze any job for improvement and then sustain any improvements we do find. Practicing Job Relations and Job Instruction on a regular basis will help create that foundation so that when the other Lean tools are activated, they can perform at their peak levels. Moreover, when we practice Lean in activities such as kaizen events, JR and JI will enable us to maintain those higher levels

---

[*] Masaaki Imai, *KAIZEN: The Key to Japan's Competitive Success,* McGraw-Hill/Irwin, New York, 1986, p. 25.

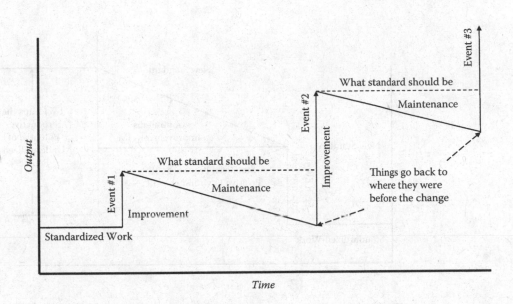

Figure 4.1   Saw-tooth improvement curve.

of performance developed in the event. Job Methods will further assist us in continuing to find improvement on a continuous basis without having to wait for the next Lean event (see Figure 4.2).

All of the case studies in this book describe the use of TWI as a driver toward achieving standard work. This is not surprising since creating and sustaining standard work, as we have just pointed out, is the foundation of Lean success. What these case studies show is that the way we are currently teaching our employees how to do their work, usually some variation of the buddy system with different bells and whistles attached, is not getting us anywhere near having operators do their work in consistent ways that follow set standards. Some companies may even deploy a train-the-trainer program that consists of, among other things, teaching trainers theoretical concepts like the different types of people and how they

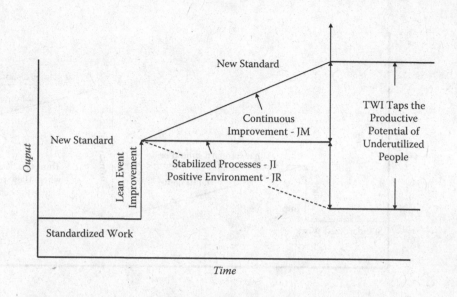

Figure 4.2 Improvement curve with Lean/TWI.

learn: auditory, visual, experiential, etc. But even in these organizations, when workers are found to be performing jobs in different ways, the typical response is, "That's the way *my* instructor taught me how to do it."

The four-step method of Job Instruction actually integrates all of these insights into how people learn by telling, showing, and illustrating the operation, and because it also includes giving the reasons why we do it that way, it ensures complete understanding. Moreover, because we take the time, before we instruct, to break down the job and put that procedure down on paper, we can be sure that the demonstration and explanation of the job is consistent every time it is taught, regardless of who is teaching the job. Instead of giving supervisors a theory on how people learn, TWI gives them a method that they can deploy every time they need to train, regardless of who they are training. It is comprehensive and works with any job we teach.

Used in conjunction with JR and JM, the JI method of training jobs is an easy-to-learn and direct method of getting people motivated to do the job following the set standards. The ensuing process stability created by the regular usage of the TWI skills will create the environment in which Lean activities overall will thrive. This was the case at Toyota when they were developing their TPS methods, and it is the case at countless companies today that are seeking to create a successful Lean experience.

SOPs vs. JIBs

If the organization is young or fairly small, it may not have any existing documentation on work procedures, and the process of creating standard work can begin anew. Companies in this situation are usually excited about implementing JI because it gives them the added benefit of having their jobs, perhaps for the first time, written down and documented in the form of a JI breakdown. This is better than having nothing at all because in order to ensure that a job is performed the same way each and every time it is done, no matter who does it, we have to have a consistent pattern, a written record, to fall back on and use when we teach and audit the work.

The fact of the matter is, though, that many companies have already gone through a process of creating SOPs for a variety of needs, including International Organization for Standardization (ISO) compliance or to fulfill auditing requirements. For these very important reasons, they must have documented the complete and correct way of doing the work with quality and safety. The SOPs are usually written by engineers or other technical people who need to instruct the factory how to build the product in the way they have designed it. For that purpose, these SOP authors, who may not work at the plant or, if they do, do not go to the production floor on a regular basis, try to include every single thing that needs to be known about or done in the job.

These detailed documents contain a wealth of good information, and instructors would be smart to study them carefully before they teach a job to an operator. Where things go wrong, though, is when supervisors try to shortcut the instruction process and simply pass on the SOPs directly to the learners, telling them to perform the job as written. After all, they assume, these documents are called the work instructions.

At one medical products manufacturer, where they had to comply with stringent Food and Drug Administration (FDA) audits, we were shown an example of a typical SOP. It was four pages long with single-spaced 10-point font type in carefully numbered paragraphs: Section 2.0, Paragraph 2.3, Item 2.3.1, Item 2.3.2, etc. Then, at the bottom of the last page, there was a signature line for the operator to affirm that he or she had read these instructions carefully and fully understood how to do the job. This was to prove, they told us, that the worker had been trained. The engineer who showed us this document, who was fairly new to the company at the time, said that even she had difficulty understanding how to do the job after having read it!

The Job Instruction Breakdown (JIB) serves a separate and distinct function from the SOP. It does not cover every single thing that happens in a job, and in that sense, it is not a written description of the job. It is made in a specific way to be notes to help organize the operation in the instructor's mind so that nothing critical to learning the job is mixed up or left out when instructing. Most importantly, the JIB sheet is *always* used in conjunction with the four-step method for Job Instruction. In Step 2 of JI, we tell, show, and illustrate the important steps of the job, as outlined on the breakdown sheet, and then do the job again stressing key points, and yet again giving the reasons for those key points, always following the breakdown sheet. But the learner is observing the job being performed even as he or she listens to these instructions. Both *what is seen* and *what is heard* comprise the full extent of the instruction.

For example, if you are changing the blade in a hacksaw, you don't need to tell the learner to "take a hold of the wing nut" or even to "turn down the wing nut." They will see you do that. All you will need to say is, "Adjust the tension on the blade," as you demonstrate how it is done. That is the most important thing that happens to understanding that part of the job.

The real skill of the JI method is to *break down* the job so that we don't give learners more information than they can handle at one time. Learning is like eating—if you try to push the whole hamburger in your mouth and swallow it at one time, you'll choke. We do the same thing when we try to teach people how to do something by telling them, in one big blast of information, everything we know about the job—or by giving them a long and detailed SOP and asking them to read and understand it. We wind up confusing them (i.e., they choke), and then they have to go on struggling to learn the job on their own, for the most part, through trial and error.

When we break down the job, on the other hand, we look for just the right amount of information to put across the job to the learners so they can learn to perform it quickly and correctly. Because they will be observing the job when they learn, the verbal instruction can be limited to just those critical aspects that would otherwise be missed. Hence, you could not just read a JI breakdown and understand how to do the job. That is why the JI breakdown sheet *cannot* act as a substitute for an SOP when, say, an outside auditor needs to read it to make sure the procedure is complete.

On the other hand, there is usually information in the JI breakdown sheet that is *not* found in the corresponding SOPs. These are usually key points concerning the knack or trick to getting the job done. When engineers take the TWI class and learn about JI, they tell us that they missed the key points in writing SOPs because they were solely focused on the procedure for performing the operation and never thought about the technique or skill it would require to get that job done. In other words, they wrote down the "what to do" of the job but neglected to consider the "how to do it." In effect, that was always left up to the operators when it came time for them to figure out how to get it done. And this is the very reason why, in spite of our emphasis on having SOPs in order to create standard work, there is so much variation in how work gets done. Every operator, or person who trains the operators, has figured out his or her own way of doing it because the technique involved in doing the jobs does make a difference. If we don't capture that in our job instruction, as represented by the key points to the job, then we never get to standard work.

What is more, SOPs tend to cover any operation in its entirety. But, on closer inspection, we usually find that the job can consist of many component parts. For example, you could have an SOP on driving an automobile, but in reality, there are many individual skills required to operate a vehicle, including steering, braking, turning, passing, parking, etc. Good instructors teach these skills one at a time so that, when they are mastered, the new operator can put them all together and skillfully carry out the entire operation. So there may be several Job Instruction breakdowns for any one SOP. This concept of dividing a job into instruction units or its teachable elements is one of the biggest insights supervisors get when learning JI. They come into the class thinking that their jobs are too large and complex to teach using a simple four-step method, but learn that this teaching method is just what they have been missing all along.

In summary, SOPs fail when we try to use them for training, but at the same time, we cannot use the JI breakdown as a substitute for the SOP. Many companies starting out with JI get so excited about the JI breakdown format that they want to use it exclusively and do away with their unwieldy SOPs that few people are using anyway because they are so difficult to understand. But when JIBs are the only written records of jobs, the temptation is to put too much information into them, and their use in instruction is thus crippled. This is what happened at Albany International (see Chapter 6) and, as a result, their initial JI implementation stalled and failed to produce the needed results. In the end, we need both documents because they serve different purposes. Table 4.1 summarizes the differences between SOPs and JIBs.

Table 4.1 Comparison of SOPs and JIBs

Standard Operating Procedures	Job Instruction Breakdowns
• Usually made by engineers or other technical people	• Made by supervisors with help from operators
• Everything that needs to be known about a job	• Do *not* contain every single thing about a job
• Used for other purposes besides training	• Always used in conjunction with JI four-step method
• Should be studied before making a JIB	• May have several JIBs for any one SOP

Controlling the Breakdown Sheets

The next big question that comes up, and it is usually a contentious one, is how to handle the JI breakdown sheets along with the SOPs, which are typically controlled by revision rules and storage requirements to ensure proper usage. Most production staff are already overwhelmed by the amount of paperwork they have to control, and trainers feel that the JI breakdowns should not be restricted by cumbersome change requirements that limit their flexibility when training a variety of people with different backgrounds and skill levels. Quality control staff may resist another version of the standard being published because, they worry, it could contradict or change their written standard. Moreover, if the JI breakdowns were not controlled, like SOPs, they would run into the same problems of, for example, outdated versions lying around and being taught to operators.

The key thing to remember here is that the JI breakdown of a job is not a different version of the job method as described by the SOP. As our good colleague Mark Sessumes from TMAC, the Texas MEP, puts it, "Standard work is not words on a page, it is behavior on the floor." In other words, the SOP itself is not the standard. The SOP describes a process that is to be done in a standard way, and the JI breakdown describes that same standard way of doing the job, but from a different point of view. It's like looking at two drawings of the same house. One may be an isometric 3D rendering of what the home looks like from the outside, whereas the other may be a schematic drawing of inside details from three different angles: front, side, and top. But both drawings still show the same house. In the same way, having another written description of the job does not mean we have another, different standard. Both documents describe the same standard process of doing the work.

Each company needs to decide how it will manage its breakdown sheets, depending on the complexity of its operations and its needs around document control. On the one hand, you don't want to become too inflexible because these are living documents that must meet the needs of a diverse and continually changing workplace—no two learners are alike. On the other hand, because our goal is, after all, standard work, we need some level of control to ensure that everyone learning the job will be given the same training and be able to perform it in the defined way. One of the best ways we have seen for controlling JI breakdown sheets is to have them managed as an addendum or attachment to the SOP that can be controlled using the same controls as the SOP. If the SOP changes, then the method of instruction must also be updated to reflect those changes. (See Chapter 8 for an excellent review of how a company with strict document controls integrated JIBs into its quality system.)

Once an operator has learned and mastered a job through proper instruction, he or she can certainly use the SOP as a reference to fall back on when questions come up or problems occur. But the need for good instruction always comes first. You should not be reading an SOP while doing a job any more than you should be driving down the highway while reviewing the owner's manual to the car. If you have a question or concern, say a warning light comes on or there is some noise or vibration with the vehicle, you can pull over to the side of the road and consult the manual. But while driving, your focus should be entirely on the task at hand—operating the vehicle safely in order to get to your destination. Without good training, you would be pulling the car over every ten minutes to check the instructions and would never get to where you're going in a reasonable amount of time.

Continuous Improvement

When searching to explain the differences in how Japanese and Western managers approach their work, Masaaki Imai, the man who taught the West the Japanese concept of kaizen, asked himself, "How do we explain the fact that while most new ideas come from the West and some of the most advanced plants, institutions, and technologies are found there, there are so many plants there that have changed little since the 1950s?"[*]

---

[*] Masaaki Imai, *Gemba Kaizen: A Commonsense, Low-Cost Approach to Management,* McGraw-Hill, New York, 1997, p. 2.

Imai explained this himself when he referenced Parkinson's Law on the natural deterioration of static systems. The U.S. automobile industry is the most visible evidence that sufficient effort was not made by industry leaders to even maintain the status quo, much less improve in response to the attack by Japanese automobile manufacturers in the 1970s. Significant strides with Lean production since the 1990s were also not enough to prevent the collapse of the industry in 2008. Based on what we have learned at companies around the world, we can't help but feel that TWI might have helped solve the tremendous people issues the American automobile industry faced, and still does, in trying to speed up the transition to Lean production.

The Role of TWI in Continuous Improvement

Bob first learned of the existence of TWI from what appeared to be an out-of-context comment by Masaaki Imai in his book *KAIZEN*: "Less well known is the fact that the suggestion system was brought to Japan about the same time [as Deming and Juran] by TWI (Training Within Industry) and the U.S. Air Force."[*] It was not until Imai published *Gemba Kaizen* years later that he credited TWI Job Methods for its role in the development of kaizen in Japan. The true value of this later book, from our perspective, was to identify the supervisor's role in *gemba* and the influence TWI had on how that role evolved.

---

[*] Masaaki Imai, *KAIZEN: The Key to Japan's Competitive Success,* McGraw-Hill, New York, 1997, p. 112.

> Frequently supervisors in *gemba* do not know exactly their responsibilities. They engage in such activities as firefighting, head counts, and achieving production quotas without regard to quality. Sometimes they don't even have daily production quotas in mind; they just try to produce as many pieces as possible while the process is under control—between the many interruptions caused by machine downtime, absenteeism, and quality problems. This situation arises when management does not clearly explain how to manage in *gemba* and has not given a precise description of supervisors' roles and accountability.[*]

That was the environment at a Nissan Motor assembly plant when Shuichi Yoshida was promoted to section manager in 1970. Yoshida later obtained a license to teach the JI, JM, and JR courses to successfully train unskilled workers to meet rapidly expanding demand in the U.S. automobile industry, much the same as during World War II. Yoshida followed the TWI formula faithfully: "Each course consisted of one week of two-hour lectures and practice sessions; after the lectures, participants returned to *gemba* to put into action what they had just learned."[†] In Yoshida's view, supervisors should be trained to act more like tutors and to look after the people they are in charge of in accordance with the roles of supervisors he defined for Nissan. Those roles are listed here, and we added the J program in parentheses that we see as having influenced Yoshida when developing each of these roles:

---

[*] Imai, *Gemba Kaizen*, p. 105.
[†] Ibid., p. 113.

- Prepare work standards (JI).
- Provide training to ensure operators do their work according to standards (JI).
- Improve status quo by improving standards (JM).
- Notice abnormalities and address immediately to keep the process under control (JM).
- Create a good working environment (JR).
- Deal with abnormalities at the worksite (JR).

Why Kaizen Is So Difficult

The key to kaizen, what we call continuous improvement in English, is that it be continuous. This seems to go without saying, but what we find in countless companies trying to install a Lean culture is that continuous improvement is not continuous. It may proceed in fits and starts, but we cannot say that it has become part of the way people in the company work every day. There are a variety of reasons for this:

- People resist change, and this mindset will cause them to backslide and abandon improvement.
- People typically rely on others for improvement.
- Even when people want to improve they don't have improvement skills.
- Companies tend to rely on scheduled events to make even small changes.
- Supervisors and operators end up leaving improvement until after "making the numbers."

The philosophy that runs through all of the TWI modules counteracts and removes these excuses. Let's look at them point by point.

We discussed earlier in this chapter how a combination of all three TWI skills helps prevent backsliding after improvements are made. But the fact that *people resist change* is a powerful force that leads to sterile and stalled Lean initiatives. The JR module directly addresses the fact that people "are all inclined to question whether it is necessary to change the things we have become accustomed to." Even in June 1944, when those words were written, people were resisting change, so this is not a new phenomenon in the age of Lean. JR then teaches that as supervisors we must prepare the way for change in the workplace and use one of the JR foundations for good relations—tell people in advance about changes that will affect them—in order to prevent problems from coming up because of the change.

The action points for this foundation are to "tell them why if possible" and to "work with them to accept the change." Telling them why helps them to understand the need for the change, and most people can agree with something, even if they don't like it, if it has a rational reason. The second point is harder to apply. JR shows us that people need time to adjust to change, and letting them "have their say" and allowing them to "blow off steam" are a part of the process of acceptance. Clarifying the effects that the change will have on individuals and explaining any benefits they will get from it also help to mitigate the fear around the upset of defined habits and routines. When supervisors don't understand this foundation and try to force change on people without walking with them through the change process, they only build up resentment and resistance.

This skill in leading is a missing piece in too many Lean implementations, even though these companies have learned that a key element to Lean success is a change in culture. We will look at this in more detail in Chapter 7, but the point here is that TWI skills add this human dimension to the Lean skill set so that we don't leave our people behind even as we press forward toward improvement.

The second point was that *people rely on others for improvement.* That means they don't feel it is their responsibility to come up with improvement ideas and suggestions. Here again, good leadership is the answer, and in JR we learn that "a supervisor gets results through people." When supervisors try to do everything themselves or when they pass the buck to others, they fail to engage their people and to "make the best use of each person's ability," another JR foundation for good relations. In the JM course, we see in the opening demonstration how a supervisor works with an operator using his suggestions in making the final improvement proposal. The TWI methods teach supervisors to conscientiously work with and through people, and this means using their innate abilities and keeping them active in the improvement process.

The next point is a situation we find much more common—people who want to participate but get frustrated because they *don't have the skills to come up with improvement proposals.* We may ask our people for their ideas, even put out a suggestion box that, after some initial complaints, remains perpetually empty. But if we don't give them a method for how to generate those ideas, they will not be able to provide them on a regular basis, no matter how much they may want to and no matter how hard we cajole them to do so. The Job Methods Improvement plan is a simple and straightforward method that everyone can participate in to find and implement improvements on a regular basis. By following the JM four-step method, anyone can come up with improved methods to doing his or her work, and it is a repeatable process he or she can do over and over again.

Though the TWI programs are directed at first-line supervisors, many companies find that the JM training course is something they want literally all of their employees to take and use. Different from JI or JR, where the skills are to teach someone else a job or to take an action on a problem concerning someone else, finding an improved method to your work using JM is something anyone can use, including those who do not give directions to others. Even when the course is given only to supervisors, these supervisors are trained to use their JM skill to direct the improvement activity while leading their people to participate in every step of the process. As the JM method states, "Work out your ideas with others," and this means getting operators involved. In order to best "sell the new method to the operators," the course teaches them to get these operators involved early in the process so that they will have more commitment to carrying it out.

Moving on to the fourth point, Figure 4.2 shows us that JM should be done continuously and not as a scheduled event. There is certainly great benefit when companies throw a lot of employee time and effort into scheduled Lean events that can then attack problems in a concerted way. But, as one frustrated Toyota executive was over-heard saying, "Kaizen is not an event, it is a way of life!" If employees have to wait for a scheduled event to make a needed improvement, they will fail to develop a men-tality of continuous improvement and will see problem solving as something that can wait until there is time to schedule it in. The philosophy of JM, as well as JI and JR, is that these skills should be practiced on a daily basis. In Chapter 10 we will describe how these skills form the basis of what is the daily management of Lean, in other words, what supervisors do every day.

This brings us to the final point, a mindset that is quite pervasive when processes are continually unstable, which is that *supervisors put off improvement until they have "made the numbers."* In the JI course, we open up by reviewing all the problems that supervisors face that hold back their production, bring down quality, and drive up cost. These are the very same problems they are wrestling with every day in order to make the numbers, and they do this through an endless effort at putting out fires, cutting corners, and pulling strings. A smarter way to meet their responsibilities of getting out quality pro-duction at the proper cost would be to develop a well-trained workforce, because we can easily see how most of the problems we face can be solved or alleviated if people are properly trained. Once good training is imple-mented and work becomes more stable and controlled, then these problems go away or are whittled down to size so that we now have time to work on improvement, which will in turn create more stability in the work at higher levels of achievement. It is not enough to simply say you don't have time for improvement. In fact, considering the fierce competition most companies face in the marketplace, they don't have time not to improve.

The true essence of Lean is to make improvement a continuous process, more a way of life than simply an activity we do from time to time. But, as we can see, this entails a lot of what has been called soft skills because we cannot simply diagram them on a chart or graph and expect them to work. In the next chapter, on the Donnelly Custom Manufacturing Company case study, we will see how a company in a tough business sector used all the TWI methods to transform its culture to one where generating improvement ideas became the way it did work on a regular basis.

Not So Japanese!

It is interesting to note that, while the Japanese word for continuous improvement, *kaizen*, is almost universally known and used not only in the Lean movement, but in society at large, the concept is not Japanese at all.* In fact, it can be found in the TWI Job Methods manual written in 1943, long before anyone considered the Japanese to be innovative manufacturers. In Step 4, when the method instructs us to "put the method to work," it also states

---

* If anything, the Japanese culture leans very heavily toward tradition and is slow to change. Before 1868, when the country was forcefully opened by Commodore Perry's "gunboat diplomacy," no foreigners were allowed into Japanese society, and it was illegal for any Japanese person to leave the country. The whole society was forced to remain stagnant for centuries by its feudal leaders. In the modern era, the Japanese had to learn how to question current methods and to embrace constant change and improvement.

that we should "use it until a better way is developed." This injunction is printed right on the pocket card that supervisors are to carry around with them at all times when they go about making improvements. The manual then emphasizes this point by stating, "Remember there will always be a better way. Keep searching for further improvements." It is completely clear here that the JM improvement plan is geared toward making a series of improvements that is continuous and never ending. This is the true spirit of kaizen.

Another well-regarded aspect of Lean that the Japanese have perfected is the use of proposal systems, or *kaizen teian* (improvement proposal) and *teian seido* (proposal system) in Japanese. Here again, the proposal was an integral part of the TWI JM program and included the use of before and after results to show the positive effects of the new method being proposed in order to help sell the proposal. Alan Robinson, in his book *Corporate Creativity*, explained the history of proposal systems that began in the United States, most notably at the National Cash Register Company of Dayton, Ohio, in the early 1900s. He states: "TWI played a central role in the development of the *kaizen teian* system. The push came from TWI's JMT course, which forcefully communicated the importance of the many small improvement ideas a company might get from its employees."*

These programs that went to Japan after the war were also described in *The Idea Book*, a Japanese book on these proposal systems:

---

* Alan G. Robinson and Sam Stern, *Corporate Creativity, How Innovation and Improvement Actually Happen*, Berrett-Koehler Publishers, San Francisco, 1997.

> The forerunner of the modern Japanese-style suggestion system undoubtedly originated in the West.... TWI (Training Within Industry), introduced to Japanese industry in 1949 by the U.S. occupation forces, had a major effect in expanding the suggestion system to involve all workers rather than just a handful of the elite. Job modification constituted a part of TWI and as foremen and supervisors taught workers how to perform job modification, they learned how to make changes and suggestions.... Many Japanese companies introduced suggestion systems to follow up on the job modification movement begun by TWI.[*]

We can see that the roots of continuous improvement lie in our own history, of which TWI was a major contributor. Even as we have learned much from the Japanese industrial experience since the end of WWII and the introduction of TWI there, it is instructive to note that these foundational elements are not Japanese at all, but have strong roots in our own culture. Before there was Lean there was TWI. And TWI was influential in the formation of what we know of today as Lean. Learning and applying these foundation elements can only enhance our use of and success in applying Lean.

Improvement as a Part of Job Instruction

As we explained at the end of the last chapter, there is an improvement element in the JI program that bolsters the kaizen effort even as we focus on developing a

---

[*] Japan Human Relations Association, *The Idea Book*, Productivity Press, Portland, OR, 1988, p. 202.

well-trained workforce. In other words, when we break down a job for training and strive to find the one best way of doing the work, all of the other not so best ways are stripped away and discarded. By standardizing the method in which people do their jobs, not only is variation removed from the work, but the performance ends up at a higher level.

Respect for People

Every company likes to claim that people, the employees of the organization, are its most important resource or asset. By taking care of that resource, the reasoning goes, the work of the company will be performed in such a good way that customers will be happy and profits will ensue. This challenge proves to be easier said than done, however, and the proof of that is easily found by talking to any one of these important resources and asking their honest opinion of their organization and how they are treated by it. It turns out that, despite the rhetoric, when it comes to how many organizations really get the work done, the usual command-and-control tools of threats and intimidation are the norm. That is, do what you are told or face the consequences.

In Section III of this book, we go into the many facets of the TWI programs that deal directly with leadership and the skills that make respect for people a reality and not just a slogan. This part of TWI's relationship to Lean is both extensive and all-encompassing. In other words, it permeates every aspect of each of the four-step methods. As we'll see in Chapter 7, the human element of the

TWI methods is what made the program so appealing to the Japanese when they were introduced to it after their devastating defeat in WWII. This philosophy of humanity in the workplace, which was so effective during the war in building up the wartime production capacity in the United States after most of the workers left their jobs to go and fight in the war, was then adopted by the Japanese as they began their march toward higher productivity and quality.

It should not be surprising, then, to find that Japanese companies operating in the United States and other countries are able to achieve this aspect of their Lean systems outside of Japan. This is no small feat considering the challenges of struggling with differences in language and culture that are very unique to the Japanese experience. Running an organization that truly respects its people and works on the company culture first, before trying to implement tools that work on the production system, is the first lesson in Lean that many, if not most, companies miss.

Since this is such a big topic, we will devote the whole of Chapter 7 to discuss it. For now, here is one story that vividly displays how the hard work of respecting people paid off for a company. We were asked to deliver TWI training for a subsidiary of Toyota in Kentucky that supplied wheels, both aluminum and steel, for the production of Toyota models at the Georgetown plant and other auto manufacturers, both Japanese and American. The plant was run identically to the main Toyota facility, so much so that when VIP guests, wanted to tour Toyota and the Georgetown plant was too busy to host them, the guests were sent to the wheel plant. When we got there, we found that several of the staff who had worked there when the plant first opened had taken the TWI Job Instruction training at the main Toyota facility when they started. But as time went on, newer supervisors were not getting the training, and so they wanted to develop the in-house ability to deliver JI and JR.

During the initial ten-hour training, one of the supervisors, we'll call Jim, approached Patrick after the class and said he had heard that one or two of the participants in the JR class were to go on to learn to become instructors of the course. He told Patrick that he would like to be one of those people because he liked the class so much. Patrick told him that it was not his decision, even though he had actually been asked by the company to submit some recommendations.

As it turns out, Jim was picked to be a trainer, and he joined us in Syracuse a few months later for the JR train-the-trainer program. It was his first trip on an airplane, and he was sure to bring his guitar with him to keep him company during the full week of training. Throughout the week, which was held at a beautiful training facility in the New York Finger Lakes, while the other participants were enjoying their free time sipping wine by the lake, Jim was found in the conference hall practicing his deliveries. He said he had to be sure he got it down just right so that he would succeed in delivering the course.

Jim told Patrick how pleased he was to be there, and said that he wanted to be a trainer so that he "could give something back to the company that had been so good to him." He told us how, when he first joined the company as a rebellious youth, he had gotten into a lot of trouble and that the company had every right to let him go. "They really should have fired me," he stated with great confidence. But the company bent over backwards to help him through this time, and Jim went on to be one of their best operators. They later promoted him to supervisor.

Jim became a very successful JR trainer, as was shown in the feedback sheets he shared with us after his first few deliveries. The people who attended his JR sessions said that he not only understood the material, but was able to apply it to the actual situation at the Kentucky plant because of his personal background. The company's efforts to develop Jim, by respecting him as an individual and working with him as he grew as a person and an employee, paid off many times over, as he could now use his own experiences to help train other supervisors to lead in the same effective ways he himself was treated.

Many executives like to make the excuse that companies in Japan have been successful because the Japanese people are better educated and motivated. But when these very same companies can replicate their success at developing a loyal and dedicated workforce in places like Kentucky, Tennessee, Ohio, Mississippi, Indiana, and Texas, we can see how these principles around respect for people are universal and necessary. You simply can't succeed with the more technical aspects of Lean until you get the people part right.

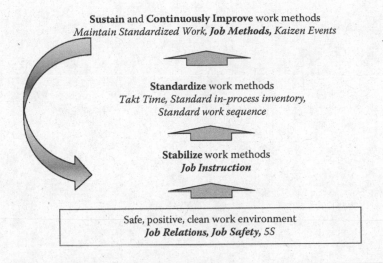

Figure 4.3   TWI/Lean integration.

TWI/Lean Integration

The fundamental progression of Lean activity is to *stabilize*, *standardize*, *sustain*, and *continuously improve*. In this chapter we have shown how the TWI programs help to build this Lean structure by directly influencing the core Lean concepts of standard work, continuous improvement, and respect for people. Figure 4.3 shows this progression and where TWI fits in with the other Lean tools. As we have said all along, TWI was there at the beginning when these Lean tools were being developed and perfected. By integrating TWI into the Lean process, by putting back into place this missing link, we can ensure that the whole system delivers on the true promise of Lean.

The Role of a TWI Champion at Donnelly Custom Manufacturing

Enterprise Minnesota (EM) is a not-for-profit economic development organization that receives financial support from the Manufacturing Extension Partnership (MEP), a part of the National Institute of Standards and Technology (NIST), which is run by the U.S. Department of Commerce. As one of the missions of these independently operated MEP centers located in every state of the United States, EM works directly with manufacturing companies to develop successful business strategies, implement state-of-the-art technologies, and smooth the transition from start-ups to mature organizations. EM has successfully leveraged this federal support for more than thirteen years by administering the cooperative agreement that supports them as the MEP center in Minnesota. In the last few years, their accomplishments include:

- Eleven hundred Minnesota companies that received critical business assistance.
- These companies realized a positive economic impact of $47.5 million to their bottom line from increased sales, reduced production costs, and improved utilization of their tangible and intangible assets, which include people.
- These companies went on to invest $39.5 million in new technology.
- More than 2,350 jobs created or retained.*

Donnelly Custom Manufacturing Company (Donnelly) is one of the Minnesota companies to have received critical business assistance from the Enterprise Minnesota regional office located in Alexandria, Minnesota. Donnelly credits the people working out of this office for providing them with consulting services that enabled them to implement Lean manufacturing concepts as a driver for their continuous improvement process. The management also credits Richard Kvasager, account manager for the Alexandria office, for introducing them to TWI, which is now playing a key role in moving their Lean program to even higher levels.

The Company

Stan Donnelly founded Donnelly Custom Manufacturing in 1984. He wanted to tap into the short-run niche market for injection molding parts that he viewed as being overlooked and underserved by many injection molding

---

* Enterprise Minnesota website, www.enterpriseminnesota.org, January 14, 2010.

companies that mainly went after the more lucrative and less troublesome long-run manufacturing jobs. Stan confidently went about assembling all of the business elements he needed for a start-up manufacturing business in Alexandria, Minnesota: a one-story building, four injection molding presses, and eight talented people. Although his new company lacked customers, he was confident that he had the right idea to focus on short-run, close-tolerance manufacturing, along with related engineering services to support this market.

The company was soon producing orders for leading original equipment manufacturers (OEMs) whose end products, like ATM machines, are not sold in mass production unit volumes but are composed of high-quality, multiple plastic components. Stan successfully built his company on the sound business principles of *speed*, *simplicity*, *service*, and *success*. He also maintained a close relationship with his people and demanded that his management do the same by always being involved in what was going on in the workplace. This approach continues to minimize debilitating communication problems other companies experience as they grow.

According to President Ron Kirscht, who joined the company in 1991:

> The best way to have committed employees is to make sure everyone has the opportunity to play an important role. All employees are encouraged to contribute ideas and suggestions to the company, regardless of seniority or position. Voluntary Lean improvement teams are formed monthly to make performance improvements on a different area of focus each time. Donnelly also has a "war room" where

cross-functional teams assemble daily to review results, identify problems and sources of waste, and then assign these to members to solve, as well as to set plans and assess progress on customer projects.

The company now has over two hundred employees who are dedicated to serving this market twenty-four hours a day, seven days each week, in a modern 110,000-square-foot manufacturing facility where the company sets the standard for short-run molding excellence as embodied in its brand promise, How Short Run Is Done™.

Understanding Short Run as a Complex Business

Ron Kirscht believes that Donnelly's respect for people is a major reason for its success in the short-run niche market. "It takes good people to deliver good products on time," he explains, "when feeding over 700 different materials through 33 presses, using more than 2,500 discrete molds that require 40 to 50 changeovers each day, to provide over 3,000 active parts for our OEM customers. Since our job security rests on performance alone, every day we go beyond the call of duty for our customers." When asked what he meant by that, Ron handed us a copy of the 2008 *Plastics News* Processor of the Year Award for which Donnelly was a finalist. In it, we found the answer.

The way to compete or succeed in this industry is to show how you are helping your clients achieve their goals and what role you play in their successes. Donnelly has received numerous supplier awards from companies that include Honeywell, Scotsman Industries,

Graco, Marvin Windows, Diebold, and Fargo Electronics, and these demonstrate that the company does in fact set the standard for short-run molding excellence. The following two examples from their award submission illustrate that excellence of service.

Scotsman Ice Systems

Scotsman is the world's largest manufacturer of ice systems, with more than three hundred models of ice machines, storage bins, and ice and water dispensers. Their equipment leads the industry in efficiency, with products such as Prodigy® Cubers and Nugget Ice® Machines that decrease both utility costs and water usage.

When Scotsman needed help with the launch of its new product, the Prodigy, Donnelly managed the complex process of developing the plastic parts for the machine from start to finish. That project involved the fabrication of forty plastic parts, the sourcing of low-cost molds to China, and finally ensuring that a high-quality product was produced and shipped on time and within budget. Donnelly's project engineer made sure that Scotsman's designs were usable and assisted their product design in ensuring the application and cosmetic requirements of the parts were met. For example, Donnelly engineering identified a tooling issue in the design stage and was able to recommend a thermally conductive mold core material that would provide adequate part cooling, improve cycle time, and eliminate part distortion concerns.

The Prodigy launched on time and on budget. "Donnelly is very attentive to Scotsman's needs and desire," said Scotsman representative Alex Harvey. "For the Prodigy, they knew what we needed and how to make that happen on time and on budget. Their partnership was indispensible." The Prodigy Ice Cuber was recognized as one of the most innovative new products in the world with the receipt of the Kitchen Innovations Award from the National Restaurant Association in 2007. According to the judges, the state-of-the-art self-monitoring system made Prodigy truly stands out.

Graco Fluid Handling Systems

Graco, a world leader in fluid handling systems and components, came to Donnelly with one of the most engineering-intensive projects they had ever seen: a $1.9 million project that required forty new molds to supply fifty-plus parts for the new Graco Husky 205 paint sprayer. Donnelly's engineering team worked up front with Graco engineers on the design of the new paint sprayer and coordinated activities with toolmakers—five domestic sources and one in China. Donnelly engineers also interfaced with prototype shops and material companies to coordinate their efforts and to keep the process moving smoothly forward. In the end, Donnelly's early involvement improved the output on this process by minimizing any necessary tooling changes, which would have delayed Graco's tight schedules.

Mastering Complexity*

Many injection molders consider shot-run manufacturing to be an unprofitable venture because of the complexity

---

* 2008 *Plastics News* Processor of the Year Award submission, Donnelly Custom Manufacturing.

of the number of parts involved in this niche market and the inherent variety of materials required. Donnelly's approach is to link its design-to-production process with unique value-added services to provide OEMs with capabilities they themselves are otherwise lacking. To achieve excellence in this market, for example, Donnelly must launch at least one hundred new molded parts in order to grow manufacturing sales by $1 million. Working closely with OEMs is the only way to keep up this frenetic pace.

The key to success in short-run molding is innovation in the mold setup process. Because of the variety of parts it produces, the company cannot purchase a dedicated quick-change mold system. The molds are simply too numerous and different, and the kinds of transfer tools needed to do the job too diverse. That led leadership to conclude that the only solution was to have well-trained and dedicated people on the job and to focus on an optimized mold changeover process.

How complex are the molding operations at Donnelly in comparison with its competitors? Michigan-based Plante & Moran is a consulting firm that helps clients analyze their operations. Their analysis calculates the complexity factor of a business by quantifying the challenges the company will face to cover overhead costs based on the number of resins, molds, and presses it uses to fulfill customer requirements. The higher the complexity factor score, the more complex the operation, and the more difficult it is to profitably manage day-to-day activities and priorities. Donnelly's current complexity factor of 45 million, as shown in Figure 5.1, is 375 times higher than a typical comparative injection molder in the industry, which dramatically highlights its effective handling

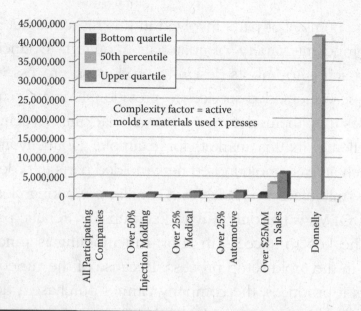

Figure 5.1   Donnelly's complexity factor (Plante and Moran's 2004 American Study of the Plastics Molding Industry). (Reprinted with permission of Donnelly Custom Manufacturing, Alexandria, MN.)

and management of the complexity of what its employees do. Its system for addressing this complexity, which focuses on accountability and a constancy of purpose, is shown in Figure 5.2.

Complexity Creates a Compelling Need for Skills Training

As is the case for most injection molders, the majority of new hires at Donnelly start out at entry-level positions. Those first hired by Stan quickly became the experts, and new employees had to watch and learn from them as they began doing their work. This buddy system approach to training enabled the company to grow in the early years, but the need for formal training soon became a priority. Future growth more and more depended on an abundance

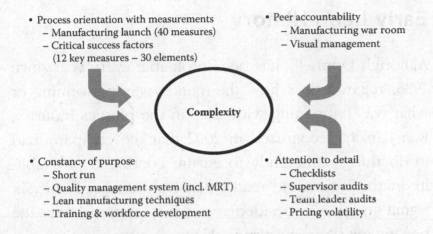

- Process orientation with measurements
  - Manufacturing launch (40 measures)
  - Critical success factors
    (12 key measures – 30 elements)

- Peer accountability
  - Manufacturing war room
  - Visual management

Complexity

- Constancy of purpose
  - Short run
  - Quality management system (incl. MRT)
  - Lean manufacturing techniques
  - Training & workforce development

- Attention to detail
  - Checklists
  - Supervisor audits
  - Team leader audits
  - Pricing volatility

Figure 5.2 Addressing the complexity factor. (Reprinted with permission of Donnelly Custom Manufacturing, Alexandria, MN.)

of well-trained operators who were capable of dealing with the increasing number, and growing complexity, of mold changeovers required for the company to continue expanding in the niche market of short-run molding.

To satisfy that need, Donnelly developed the Donnelly Certified Molding Operator (CMO) Training Program in 1991. A combination of classroom and vigorous hands-on training, the program is one where operators must apply to get into, be recommended by their supervisor, and be approved by manufacturing leadership, including President Ron Kirscht. Certification is achieved by passing a written test and by demonstrating ability to perform challenging changeovers to standard. Upon certification, new CMOs receive a bonus, a raise, and a "Certified Molding Operator" shirt, and their name is added to the CMO recognition plaque on display in the manufacturing area. Donnelly currently has eighty-four CMOs who comprise 40% of the total workforce. As the company continues to grow, the position of CMO is seen as a promotion for those eager enough to attain the distinction.

Early Lean History

Although Donnelly has been profitable every year since 1986, regardless of how the market was performing or what was happening with sales in the plastics industry, Ron Kirscht recognized in 2002 that the company had to do things differently to ensure continued profitability in the future. He read extensively on Lean and Six Sigma and was undecided as to which approach was the best fit for their short-run, close-tolerance plastic injection molding business. Ron decided to bring in two consultants to conduct a side-by-side trial of Six Sigma and Lean to evaluate both before making a decision. As for the Six Sigma group, by the time the data were collected, the situation being studied had changed and the solution to the trial problem no longer applied. This led the GE representatives facilitating the process to conclude that Six Sigma did not fit with the Donnelly short-run business model.

As for the Lean trial, Lean 101 training was arranged through Enterprise Minnesota for all employees in December 2002 to introduce them to the fundamental concepts and tools of Lean. This full-day hands-on assembly simulation involved groups of employees learning and applying the tools of Lean to transition a traditional manufacturing assembly operation into a Lean process. Having participated in this training, and seeing the applicability of the concepts to Donnelly's operation, Ron knew he was headed in the right direction with Lean.

A 5S event was conducted in January 2003 to demonstrate how having a neat, clean, and safe work area can solve workplace problems that employees had learned to live with over the years. The initial training was then followed by a series of eleven Lean events in 2003 to create and install improvements in the plant. Five more events were conducted in 2004 before the improvement program stalled. While the early results looked so promising, the effort ran out of gas as "low-hanging fruit" proposals were quickly picked off and the Lean events devolved into just another routine task to do and get out of the way. They were not able to sustain the initiative.

Ron, who had continued studying Lean as he observed what was happening to the Lean initiative on the floor, recognized that this tactical approach to the effort was not generating the expected results. It was just a bunch of tools that were not becoming a part of the culture. He decided to start over by involving Donnelly's management team in a strategic planning process that, among other things, identified the need for an experienced professional from outside of the company to coordinate and drive the change process on a full-time basis.

Search for a Lean/TWI Champion

Ron approached Sam Wagner and asked him if he would like to come on board with Donnelly to lead the Lean effort. Sam was someone he had come to know when they both served on a volunteer board for Douglas Country Developers, a community action group that supported economic development activities in the Alexandria area. Sam welcomed the challenge and joined Donnelly in October 2004 as the director of advanced manufacturing with four direct reports: the director of manufacturing, the quality manager, the training and continuous improvement coordinator, and the process engineer. Having already learned from past experience that doing Lean events alone was not going to get Donnelly where management needed the company to be, Sam came on board and collaborated with Ron to develop a plan to reinvigorate their Lean program by tapping into the creativity of each of the company's workers.

Born and raised in Detroit Lakes, Minnesota, Sam Wagner put his engineering degree to work at General Dynamics in San Diego, where he earned his MBA before moving his young family back to the Alexandria, Minnesota, area that he and his wife, Cindy, considered their home. After spending two years as a regional director for Enterprise Minnesota, Sam went back to his manufacturing roots as the plant manager for the Gardner Bender's Palmer Industries, where he became a believer in Lean manufacturing while working with the Kaizen Institute to implement Lean manufacturing at this plant. Sam made an untimely move to become president and general manager of an automotive frame straightening equipment manufacturer in Alexandria that was soon closed due to corporate downsizing. Not wanting to move his family, he became a production supervisor for Superior Industries, where his immediate challenge was to train people assigned to the newly created position of team leader.

In search of training for this group, Sam contacted Rick Kvasager and Bill Martinson, who were former colleagues of his at Enterprise Minnesota. They enthusiastically explained that TWI was the answer he was looking for. Sam immediately arranged to have his people attend a ten-hour JR training program in 2004, where they learned the JR four-step method on how to get results through people. The feedback from the team leaders was fantastic. According to Sam, "They referred to JR as being the best training they had ever received." Sam followed up the JR training by meeting with the team leaders on a biweekly basis to standardize how company policy was to be implemented consistently throughout the plant. This initiative was such a success that the experience turned Sam into a staunch TWI advocate shortly before joining Donnelly.

The Lean/TWI Journey

Sam joined Donnelly knowing that management understood the importance of sustaining new initiatives, rather than continually starting a new flavor of the month program that Ron referred to as *Whim du Jour* management. With the help of Rick and Bill at Enterprise Minnesota, TWI quickly became a key part of the Donnelly Lean implementation plan. Since the other people reporting to Sam on the leadership team had day-to-day operating responsibilities, it was decided to have these three team members become TWI trainers. JI was a natural extension for Bradley (Brad) Andrist, who was already responsible for training as the training and continuous improvement coordinator. As the director of manufacturing, David Lamb was addressing execution and personnel issues in the workplace, so he opted for the JR module. Sam Wagner, director of advanced manufacturing, viewed JM as a foundation for an idea program to engage people in the improvement process and was the module he wanted to direct. Donnelly now had an enthusiastic cadre of people, supported by inspired leadership from the top of the company, ready to move forward with TWI—a formula that would ensure a high level of success with sustained commitment.

Introducing TWI in Minnesota

With funding from the West Central Initiative, Enterprise Minnesota arranged for people from Donnelly and another local manufacturer, TWF Industries, along with some of their own field engineers from Enterprise Minnesota, to participate in the ten-hour supervisor program for JI and JM training. The training was successfully conducted in Alexandria in September 2005 at the Donnelly plant. All three participating organizations followed up this initial training by having at least one of their people trained as instructors in Job Instruction in November 2005 so they could begin conducting the JI class on their own. Since then, Enterprise Minnesota has developed five TWI Institute-certified trainers who deliver the TWI programs of JI, JR, JM, and Job Safety (JS) for companies throughout the state as part of their NIST MEP Lean program.

Rick Kvasager credits Donnelly for "opening the door for TWI at other Minnesota companies." He explains further:

> Hardly a week goes by that we aren't providing TWI training for a company that wants this training because others see immediate results. Whether it's freeing up management time from the skills learned in Job Relations, standardizing the way people are performing their tasks in Job Instruction, or money saved through improvements made with Job

Methods, TWI training will typically pay for itself by the end of the first week of training.

Implementing JI

Brad Andrist, the fourth of eight people first hired by Stan Donnelly in 1984, was brought on as a press operator after receiving his degree in fluid power from the Alexandria Technical College. During his first nineteen years at Donnelly, Brad progressively moved up from press operator to supervisor, production manager/ maintenance/production scheduler, corporate trainer/ production scheduler, assembly manager, production manager, and then to his current position of training and continuous improvement coordinator. His background and his diversified experience at Donnelly were ideal for him to become the company's JI trainer. Brad enjoys his work at Donnelly, which enables him to live on a hobby farm just outside of Alexandria with his wife and two children, where he also enjoys fishing, hunting, and snowmobiling when not working on cars.

Brad's initial JI training exceeded his expectations. Ron was also very pleased with the initial results. As Ron explained it, "Implementing the new TWI training methods at Donnelly in 2005 resulted in a dramatic cut in training time, from months to days, while employee turnover rates continue to be low. Feedback from employees told us that the TWI training made their jobs less stressful." Using JI, Donnelly soon uncovered opportunities for improvement in other areas that had an immediate need of JI training. This led the company to modify and slow down its original implementation plan so it could take advantage of these opportunities while building momentum for the program plant-wide.

In one of Brad's early classes a new employee asked if everyone was trained on how to properly run an injection molding machine, the most common job in the whole plant. Unlike changing a mold, a complicated and involved process, this was considered to be common knowledge by those who do the job every day. Trainers had not even considered that a new employee would not learn and thus know how to perform the basic tasks by watching other operators. When he looked into the matter, Brad was actually surprised to learn that they didn't have any written standard instructions on how to run a machine. A team then applied the JI method to create a JI breakdown on how to run an injection molding machine, and now every new hire in manufacturing is trained on running a machine using that standard. "That problem went away," says Brad, "now that every employee is trained using the JI method that requires a person to perform the job with someone observing and coaching until that person demonstrates that he or she knows how to properly operate an injection molding machine."

JI Finds a Home

Follow-up audits have been critical to the success of the JI implementation because they opened up the lines of communication between the trainer and the front-line operators. This is very important at Donnelly, where, when running forty to fifty different and complex jobs, every day is different and many of the jobs are not run each day; some jobs are only run once a week, while others are run with even longer intervals in between, causing operators to request retraining when these jobs are scheduled. Brad discussed this with the operators, and they decided to keep JI breakdowns at each press for operators to use themselves as refresher training on how to run a job they had not done for some time. This not only solved the retraining problem, but operators got more job satisfaction by having the responsibility to "train themselves" rather than having to ask for assistance.

Today at Donnelly, using the TWI JI training methodology requires significantly less time than the traditional training methods it replaced, with significantly better results. For example, in 2007, Donnelly suddenly lost a number of the people who performed changeovers on one shift, and this created a void that put the company at risk of not meeting all customer delivery requirements. The supervisor on the impacted shift estimated that it would take a minimum of three months to get his shift back to normal. Donnelly knew it had to react much more quickly than that. Sam, Dave, and Brad put their collective TWI heads together and developed a plan to put JI to the test to speed up the training process:

1. Update the pertinent changeover job instructions (JI breakdown sheets).
2. Use the new job instructions to retrain setup operators on how to train other operators.
3. Coach newly trained setup operators on how to use these job instructions to train new hires.

After just two weeks the supervisor reported that one-half of the training was completed. This worked so well that the process was soon implemented throughout the company to get new people up to standard, reducing training time by two-thirds.

Creating an ISO 9001 Control for Job Instructions

Until the company became ISO 9001 certified in early 1996, Donnelly did not have any formal documentation with which to train people. Even after that, like at most ISO-certified companies, no one at Donnelly realized just how poor their ISO work instructions were at training people until it was discovered that copies of these work instructions, stored in the office, were seldom, if ever, used in the workplace. The work instructions were then placed on the shop floor where, again, they were underutilized by trainers who reported "they are too long to train from."

JI breakdowns, or JIs, as they are referred to at Donnelly, have since been created to train people how to perform job-related tasks in compliance with ISO's work instructions and standard work rules. The supervisor of a process is now responsible to revise the JI when a job changes, or as soon as a new piece of equipment comes in the door. The supervisor assigns a person or a team of people most familiar with the job or the piece of equipment to make a rough draft of a JI breakdown. Brad then reviews it before sending it to document control to be assigned a control number. The new JI is then placed on the piece of equipment where it is used to train all operators, and also as a refresher for any trained operator who has not recently operated the equipment.

In accordance with the ISO 9001 standards, all processes at Donnelly are to be performed in accordance with the latest issue of the applicable work instructions. New documents and changes to existing documents follow the same updating process as all controlled documents. Training documents are given a separate document title that is controlled by the revision number and letter designator. These are fed into Donnelly's electronic training database system (IQMS) to track every employee training need by skill along with training received. This training database includes all current and past employees, their current skill sets, and training items needed per skill set, as shown in Figures 5.3 and 5.4.

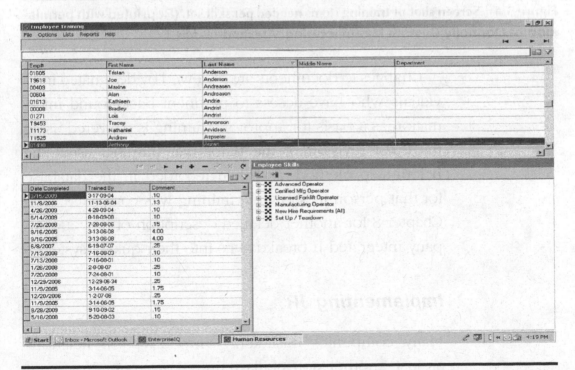

Figure 5.3   Screen shot of current skill sets. (Reprinted with permission of Donnelly Custom Manufacturing, Alexandria, MN.)

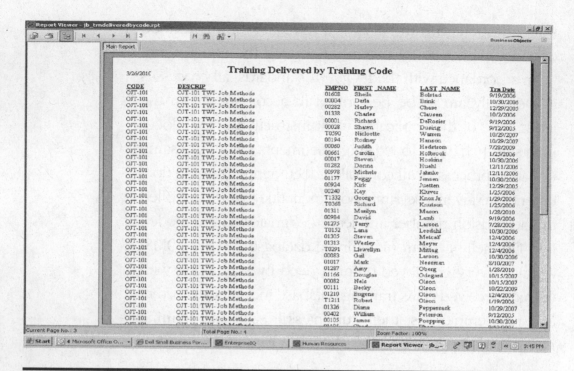

Figure 5.4   Screen shot of training items needed per skill set. (Reprinted with permission of Donnelly Custom Manufacturing, Alexandria, MN.)

Changes to documents are reviewed by designated individuals who have access to pertinent background information as a basis for approval. Training is delivered using the latest revision of the JI for that training, using the JI four-step method, and then recorded in the database for that person to show the training was completed. (See Chapter 8 for another detailed description of how a company integrated JI breakdowns into their quality system.)

Implementing JR

Many companies mistakenly view TWI Job Relations as just another soft skills training program. Not so at

Donnelly, where David Lamb, director of manufacturing, immediately recognized how JR would fit into the leadership training the company has been providing since 1993. "Both programs stress the need for supervisors to work through problems with people," David says, "but JR has a much broader definition of a supervisor that goes beyond the traditional superior-subordinate relationship as being *anyone* who gives work direction. I was excited about being the JR trainer since the most difficult personnel issues that crop up in the production area end up in my office. I could see that having me provide JR training to the people on our production floor would not only improve relationships in the workplace, but would also make my job a lot easier."

Dave earned his BS degree in industrial technology before becoming a production supervisor at Texas Instruments where he broadened his education earning a BA in economics before moving on to work for Toro. Although Dave started out as a production supervisor, he soon advanced to more responsible positions as molding manager, total quality facilitator, and advanced manufacturing planner. His interest in working for a smaller company led Dave to join Donnelly in 1995 as their tooling coordinator, where, after one year, he was promoted to production manager and then to his current position as director of manufacturing.

Dave's initial focus for the JR training at Donnelly was on the managers, supervisors, and team leaders before expanding the training to include other employees. Dave soon realized that, like all the TWI modules, learning occurred as each participant brought a real-life problem to solve in class using the four-step method learned in the first day or two of training. He explains:

> What struck me early on was how people tended to jump right to "weigh and decide" or "take action" before gathering all of the facts by talking to all of the people involved. This created the same poor results as in the case studies analyzed in the first two days of the ten-hour class. More than 80% of the problems that came to me after the initial training were easily solved when the supervisors were coached to realize they did not have all of the facts before taking action. It was also apparent that when it comes to dealing with people problems, we usually do not stop to consider what our objective is before moving forward to solve a problem, a practice that I must admit to being guilty of myself.

JR Finds a Home

Dave is pleased that being a JR trainer does in fact help him to do his job better, especially when he unexpectedly runs into a problem. "Personnel issues come my way on a regular basis," he says, "and the issues that now come to me are increasingly resolved before they have time to grow into bigger problems, making them easier to resolve." Dave attributes this big step in the right direction to his supervisors and team leaders practicing the JR foundations for good relations and learning to treat people as individuals. He believes that his supervisors are increasingly taking the emotion out of the problems they face by following Step 1 of the JR four-step method,

Get the Facts, and the caution point for that step, "be sure you have the whole story," before moving to Step 2. This is just what occurred in the following situation Dave relates that happened in the recent past:

> A very new employee was observed abusing the company's cell phone policy after having been informed of the policy, warned, and written up in the short time this person worked at Donnelly. The supervisor who observed this person skipped Step 1, "Get the facts," and went right to weighing, deciding, and taking action by recommending that this person be fired. When the problem was brought to my attention, I pulled out my JR pocket card and walked the supervisor through the four-step JR method. In doing so, we discovered that the person charged with this cell phone offense was actually not the same person who had been warned on this policy in the past. Further investigation revealed that the employee took the call only because of a family emergency.

The above example demonstrates how quickly the potential firing of an employee became a positive learning experience for that supervisor, who will certainly change the way similar issues are approached and handled in the future. The employee also gained respect for the company by being treated fairly over an issue that could have been handled in an unfavorable manner. These kinds of misunderstandings don't always have a happy ending, so Dave made it one of his priorities to follow up with all of the supervisors to be sure they understood the importance of getting all of the facts. He also took this opportunity to review the other side of the yellow JR pocket card—how to prevent problems from happening in the first place by treating people as individuals.

Measuring the Impact of JR

Donnelly has been conducting annual employee attitude surveys since 2003 that include questions on safety, quality, fairness, and corporate values. The results of these surveys have improved every year since JR training was introduced, but these results were improving even before they started JR training, so it was difficult for them to state that the improvements were due to JR training alone. "What I do know," says Dave, "is that fewer personnel issues reach my desk now than in the years prior to JR training. I have more time to devote to managing change in the workplace now that most issues are quickly resolved at the supervisor and team leader level in our organizations. There is no doubt in my mind that is due primarily to the JR training."

Implementing JM

Sam Wagner published an article in the February 2009 issue of *Mechanical Engineering* in which he wrote, "The success of any company's Lean journey relies on the willingness of the workforce to embrace change by adopting Lean principles. Training Within Industry makes it easier for a company to attain the much-desired goal of getting everyone to do their jobs correctly and efficiently." Since Donnelly's short-run business has a very short value stream—forty to fifty unique jobs run daily—different problems come up every day, and engineers cannot be expected to solve all of the problems that come up in the workplace on a daily basis. These circumstances led management to believe that JM would fit in perfectly at Donnelly.

Knowing that his initial challenge was to get supervisors on board with JM before moving forward with plans for an idea program, Sam first presented JM to training managers, supervisors, and team leaders as having already proven itself to be the best approach when Job Relations was introduced. Being able to teach and put his people through the JM course proved to be just the right vehicle for getting the supervisors on board, as the following story Sam tells suggests:

> One of the two persons who volunteered at the end of the first session to bring a problem to solve in class came unprepared because, according to that person, he had "walked the floor from one end to the other and could not find a job that could be improved." Unable to accept this as a reason, I asked this participant to randomly select a press and I would take the class out on the floor to find improvements. My suspicion that this was a test for TWI proved to be correct when the most streamlined automatic press in our plant was selected.
>
> We listed every detail on a JM breakdown sheet as I walked the class through the job one detail at a time. We then questioned each detail using the why, what, where, when, who, and how sequence to determine whether details could be eliminated, combined, rearranged, or simplified, just as I had done with the microwave shield demonstration in the JM class the day before. One of details was for

the operator to "lift the transfer box onto the table." Questioning why this was necessary led to the idea that it would not be necessary to lift the transfer box onto the table if the parts were allowed to fall directly into a finished goods container. Many more improvements came out of this exercise, and this proved to be the beginning of an attitude change for the entire group.

The participant who initially said he could not find any areas for improvement commented afterwards that the best part of the training was "shattering the box I was functioning in, and I usually pride myself on being an out-of-the-box thinker!"

JM Finds a Home

Sam experienced his own out-of-the-box experience when attending the TWI Summit in the spring of 2008. Although Donnelly had started Job Methods training in September 2005, it wasn't until December 2006 that 20% of the workforce was trained. Though they started to get some traction in terms of consistently receiving a few improvements every month, Sam knew they could do better. At the TWI Summit, he heard Dr. Alan Robinson speak about workforce idea systems[*] and suddenly visualized how the JM process could be used as an effective idea system to better engage the workforce. He began working with the shift supervisors, asking them how they could get more ideas from their people, and that led them to simplifying the original JM proposal

---

[*] Alan G. Robinson and Dean M. Schroeder, *Ideas Are Free: How the Idea Revolution Is Liberating People and Transforming Organizations*, Berrett-Koehler Publishers, San Francisco, 2004.

sheet (see Figure 5.5) to better fit the types of improvements that they were looking for. As Sam explains, "The ideas started flowing in as we increased the number of employees trained to 40% in September 2008."

Comfortable that he was on the right track for a workforce idea system, Sam worked with Ron, Dave, and the supervisors to set a goal for each supervisor of at least one improvement per shift per week. A visual management system was set up to track the progress (see Figure 5.6). Different approaches were tried by different supervisors to achieve the goal. Some supervisors gave people time to do a JM improvement during working time rather than running their machines. Others simply told their people that they were expected to submit ideas while performing their normal work tasks. Both ways worked. The supervisors reminded people of improvement ideas they had already had in the past and encouraged them to write them up since it was their idea to begin with. As people began reading the communication board where JMs were posted, they began to learn what others were doing, and Sam encouraged them to adopt ideas from other areas and shifts and to give those areas credit for submitting an idea to another functional area.

By April 2009 ideas were being submitted at a rate of over one hundred JMs, as they called the Job Methods proposal sheets, per month. As employees began to exceed the goal of one idea per shift per week (the line on the bar graph in Figure 5.7), the leadership team raised the bar to a new goal of two ideas per shift per week.

Submitted to: _____ Made by: _____

Employee number: _____

Mold number: _____ Department: _____

Product/part: _____ *Date tested: _____

Operations: _____

The following are proposed improvements on the above operations.

1. Summary:

We worked on improving the following process:

Our analysis shows the following problems:

To improve the process we did the following:

2. Results

	Before Improvement		After Improvement	
Cycle time (seconds)				
Operator time needed per cycle (seconds)				
Amount of walking (steps)				
Reject rate				
Number of operators				
Quality				
Safety				
Other				

3. Content

Description of changes and benefits:

Thanks to those who helped:

Was
Updated:

Process/picker/robot sheet updated and scanned?   Y or N

Operator instructions updated and copy attached?   Y or N   Do not scan

Special/auxiliary setup instructions updated and copy attached?   Y or N   Do not scan

Mold requires Sprue picker (emailed Lee)?   Y or N

If proposal: Does a RFA or maintenance WO need to be completed?   Y or N   Attach original to this JM

Route to supervisor for approval before implementing:

_____   _____
Signature   Date

(*Supervisor only*) This can be implemented: Yes or no and why? _____

Staging information?   Y or N

Does the BOM need to change?   Y or N

Then does a DCO need to be completed?   Y or N

Note: Explain exactly how this improvement was made. If necessary, attach present and proposed breakdown sheets, diagrams, and any other related items.

If improvement has not been tested, write *proposal* on "date tested" line.

Figure 5.5 Donnelly Job Methods proposal sheet. (Reprinted with permission of Donnelly Custom Manufacturing, Alexandria, MN.)

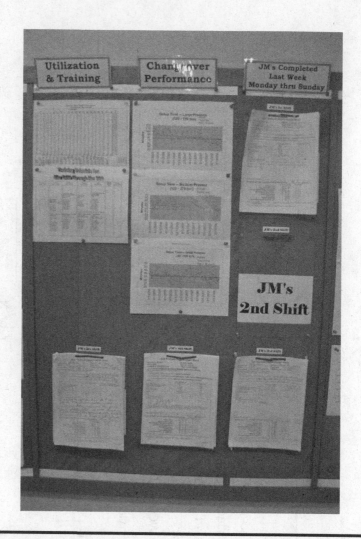

Figure 5.6 Visual management system—improvement using JM. (Reprinted with permission of Donnelly Custom Manufacturing, Alexandria, MN.)

Employees responded with even more ideas. When their supervisors expressed concern that employees would run out of ideas, Sam provided them with a copy of an article* that explains how companies with great workforce idea systems never run out of ideas, and why.

---

* "Toyota's Idea Factory," excerpted exclusively for *Electrifying Times* from Alan G. Robinson and Dean M. Schroeder's 2004 release of *Ideas Are Free*, www.electrifyingtimes.com/Toyota_idea_factory.htm.

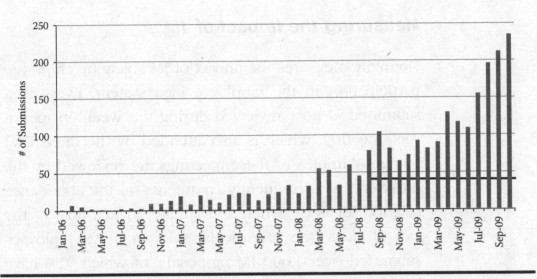

Figure 5.7 Job Methods improvement activity. (Reprinted with permission of Donnelly Custom Manufacturing, Alexandria, MN.)

Lessons Learned about Idea Programs

Sam learned a great deal from this experience in terms of the true potential of the workforce to participate in and contribute to a truly continuous improvement system. Even though he had entered this job with a deep understanding of Lean and the kaizen process, he was still dumbfounded at the contribution TWI Job Methods could make to integrating people's ideas into the process. The following points sum up his learning:

■ Employees have lots of ideas—be ready to deal with what you ask for.
■ Build your system to fit your structure and culture.
■ The system must be simple to use. The fewer the approvals and the less paperwork, the better.
■ Train supervisors on how to encourage ideas.

Measuring the Impact of JM

Donnelly measures the impact of JM solely by employee participation in the employee idea system. Every idea submitted is now reviewed during the weekly supervisors' meeting, which is also attended by the three TWI trainers. Minutes of these meetings are reviewed by the president, who frequently comments on the abundance of good ideas or, when appropriate, makes note of the lack of ideas from a particular shift. In 2009, employees submitted over 1,600 JM proposals, of which 97% have been implemented, with a participation rate of over 80% of the workforce. The company looks forward to breaking this record in 2010.

Evaluation of Their TWI Implementation

Based on these efforts, all three modules of TWI were successfully implemented into the Donnelly working environment, and the company could fully take advantage of the new skills its people developed in advancing the Lean program. Because the TWI activities they performed on a daily basis fit in so well with what the company was trying to become, they soon became part of the company's culture to the point where, if someone didn't follow the TWI prescribed method of instructing, solving a personnel problem, or improving a job method, they were called out on it. In other words, the TWI methods became the expected way that these leadership qualities were to be carried out.

Nevertheless, in looking back at how they got TWI embedded into the company's culture, while there were many key elements that led to the success of the implementation, there were also things they wish they could do over. Here is a bulleted summary list of the things they felt went right and the things that they feel could have been done better.

What They Did Right

- Selecting three members of the leadership team to become TWI trainers, which sent the message that TWI was a priority for the company.
- Placing TWI on the agenda for the leadership team's weekly progress meetings with the president, who demonstrated a constancy of purpose for the initiative.
- Developing in-house TWI trainers who have production floor experience and a good understanding of the Lean tools and concepts. This is needed to help facilitate the TWI method of solving actual problems in class.
- Conducting trainer follow-up sessions with people in the workplace to make the training stick.
- Seeking assistance from Enterprise Minnesota to network with other companies in the area that were dealing with similar challenges when implementing TWI with Lean.
- Having the leadership team work with employees and with customers to solve Donnelly-unique questions that

come up during a training session so these questions can be addressed in other TWI training sessions.

■ Persistently evaluating how to make TWI training more effective for their operation.

■ Attending the TWI Summit each year to learn best practices and how other companies have overcome challenges they experienced when rolling out TWI.

What They Would Do Differently

■ Initially introducing all three 10-hour programs of JR, JI, and JM within a six-week period was too aggressive to properly manage.

■ After the initial introduction of the program, being too slow in scheduling the classes they needed to get more people trained in the TWI methods.

■ Focusing the first ten-hour training on supervisors, support personnel, team leaders, and buddy trainers whose goal was to create Job Instruction breakdowns. They realized too late that they really needed more people trained up front to acquire and start implementing JI training, above and beyond just getting breakdowns written.

■ Thinking TWI could be a series of ten-hour classes. Follow-up is required for each J program to take hold, so that old habits are broken and new ones take their place. This takes a lot of focused time, energy, and persistence on the part of supervision that was not taken into account up front.

■ Thinking they could use JM to identify the one best way to run a job without having a full-time press operator involved. They quickly learned that every job is run differently, depending on which press it is run on and other factors. This taught them to create a JM breakdown as the operator demonstrated the current best method, so they could analyze one step of the process at a time to agree upon the improved method to standardize.

Putting TWI to the Test

Once they had TWI under way and were seeing good results across the board, the leadership team ran into a test to see whether these methods could really help them improve performance to the levels they needed to meet increasing business demands. They had to see whether these newly ingrained skills could prove them selves under fire when the company was called upon to do even better than they felt was their best.

Though the universal measure of quality is defective parts per million (PPM), Donnelly felt this was not a good measurement in a business where production runs will range from 25 to 100,000 parts per order. In a business model that requires multiple mold changes involving different materials daily on each machine, they believed that using cost of quality as a percentage of sales was a better gauge of their performance and began measuring this in 1996 (see Figure 5.8). The internal cost of quality for fiscal year 2008 was 0.34%, down from .049% in 2007. This means that less than one-half of 1% of the company's production value had to be reworked or scrapped.

The people at Donnelly thought that these numbers were pretty impressive, especially when you consider

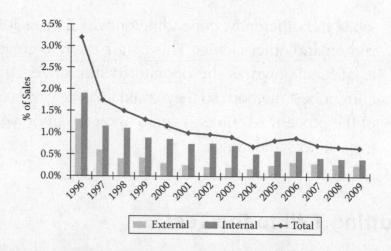

Figure 5.8 **Total cost of quality performance. (Reprinted with permission of Donnelly Custom Manufacturing, Alexandria, MN.)**

the complexity factor involved with short-run manufacturing. However, close to zero defects is no longer enough in today's world. When Donnelly began measuring quality PPM in 2007, they ended that year with a defective performance of 2,526 PPM that was not good enough for their large OEM customers, who have come to expect zero defects from both their low- and high-volume product manufacturers. Donnelly got a wake-up call when one of their large OEM customers rejected an entire shipment because of one human error. This rejected shipment resulted in a PPM of 1,341 vs. a customer expectation of 500, and Donnelly was put on notice to take corrective action.

Sam researched the subject of human error, and after reviewing various options on how to reduce mistakes, he developed a structured approach to mistake-proofing in the form of a cross-functional workshop based on the proven TWI structure and learn-by-doing model. The plan went as follows:

■ The workshop is conducted in two-hour segments over four to five consecutive days.

■ It consists of small classes of eight to ten people.

■ Each participant demonstrates and learns from the others.

■ Participants are to bring in current real problems from areas within their control.

■ They are to create a pocket-sized card with the four-step method modeled after the JR card.

Sam and his team actually created, on their own, a four-step method for problem solving that integrated the TWI skills they already had. The method is shown in Figure 5.9. It was only later Sam learned that, after World War II, when TWI was being introduced into Japan, the

---

Identify what went wrong.

Step 1. What mistake was made?

Step 2. Identify the root cause(s).
Has the job been simplified using Job Methods?
Was Job Instruction used to train effectively?

Step 3. Develop the best solution.
Consider cost vs. risk.
Is the right way the easy way?

Step 4. Apply the solution.
Check to make sure the risk has been resolved.

Figure 5.9 Donnelly's four steps to problem solving. (Reprinted with permission of Donnelly Custom Manufacturing, Alexandria, MN.)

same concept had been arrived at and a TWI problem-solving program was developed and implemented (see Chapter 12). As of this writing, Donnelly is planning a trial delivery of the TWI PS program, which was updated and put back into practice in the United States by the TWI Institute in 2009.

The Results Were Immediate

Using the TWI four-step methods, as well as the training skills they learned from doing the TWI classes just as they were developed in the 1940s and used in Japan since then, Donnelly was able to leverage its abilities to resolve quality problems and get back on track with its quality production. The results were both immediate and impressive. They were able to substantially reduce rejects due to mistakes from 2,526 PPM in 2007 to 853 PPM in 2008. What is more, they were able to change the mindset of the workforce from a "just try harder" mentality to one where they began seeking out and finding more permanent fixes to their problems. The result was that there was much less finger pointing and more celebrating the successes of the problem-solving teams.

The proof of their success, though, was demonstrated when their improved quality performance led them to becoming the top supplier for the major OEM customer that initially called into question their performance. They had proven that they had what it takes to stay competitive and to stay on top. In addition to this success, they were also, through these efforts to improve quality, able to reduce reoccurring damage to the machines, which further added to the bottom line achieved.

In a recent article published in *Quality Progress*, Ron Kirscht pointed out: "A great lesson of TWI is the process itself. It can be modified into a process that is uniquely yours. At Donnelly, we took what we've learned from TWI and created another quality and training process centered on mistake-proofing."*

The Impact of Lean/TWI on Sales, Income, and People

Lean was introduced to Donnelly in 2003, followed by TWI in 2005 and both programs were fully integrated in 2007. Annual sales increased over 57% during these four years, an average of 12% each year. Income increased by 200% during this time period, allowing the company to invest in people (including company-wide bonuses, which increased by over 250% during this same period), facilities, and new technology to better serve customer needs and maintain its competitive edge. There is no question that the combination of Lean and TWI at Donnelly contributed immensely to these outstanding performance results.

Like most other manufacturers around the world, Donnelly experienced a sudden reduction in orders from all of its customers as a result of the economic crisis that surfaced globally in the last quarter of 2008. President Ron Kirscht informed us that although the company recorded a slight increase in sales and profits in 2008, results were negatively impacted by the sudden drop of

---

* Ron Kirscht, *Quality Progress*, Assets, January 2010, p. 38.

business and are therefore not included in their year-to-year comparisons. While they had to lay off 10% of the workforce in early 2009, in April of that year they went on Minnesota's Workshare Program, which allows a company to "lay off" employees for one day a week while maintaining their employment for the remainder of the week (essentially a 20% reduction). This enabled Donnelly to retain their trained and talented workforce while reducing overall expenses. In September, they were able to leave the Workshare Program and began to recall the employees that were laid off earlier in the year. By December 2009, they had recalled everyone and began hiring additional people once again.

According to the Plante Moran 2009 Report, which was based on 179 North American plastics processors and published in *Plastic News* on February 25, 2010, median employee turnover in the industry was 34.4% for all companies and 25.5% for successful companies. Donnelly's turnover was at 10.2% for that same time period. While there are many factors that come into play here, these figures clearly indicate that people view Donnelly as a good place to work thanks to the many efforts of management to engage people in their work.

As Peter Drucker wrote: "Work belongs to the realm of objects. It has its own impersonal logic. But working belongs to the realm of man. It has dynamics. Yet the manager always has to manage both work and working. He has to make work productive and the worker achieving. He has to integrate work and the worker."* Donnelly appeared to be going down that path.

---

* Peter F. Drucker, *Management: Tasks, Responsibilities, Practices,* Harper & Row, Publishers, New York, 1973–1974, p. 169.

Closing Remarks from President Ron Kirscht

Ron attributes the company's success to the culture at Donnelly, which is focused on improving customer satisfaction and fighting a continuous war on waste. He explained the role TWI has played in the development of that culture:

> The focus of the war on waste is to eliminate wasted machine time, wasted labor time (value-added per employee), wasted materials (scrap), and the biggest waste of all—people's belief systems. As a company, we have learned much since starting down the path to Lean in 2003, but there is much more to learn. One of the biggest lessons we have learned is that the success of any company's Lean journey relies on the willingness of the workforce to embrace change.
>
> Donnelly has long recognized the value of proper workforce training as a key component of our success in short-run injection molding. Traditional training methods take too long, and employees didn't always retain knowledge learned in the process. Training Within Industry (TWI) has made it easier for us to attain the much-desired goal of giving people the skills they need to perform their work correctly, safely, and conscientiously in conformance with the company's commitment to continuous improvement and customer satisfaction. As Pete Gritton, vice president of HR Toyota North America, wrote, "Moving machines takes minutes, but changing the way people think and act takes years."*

---

* Jeffrey K. Liker and Michael Hoseus, *Toyota Culture: The Heart and Soul of the TOYOTA WAY*, McGraw Hill, New York, 2008, pp. xxi–xxii of the foreword by Pete Gritton, vice president of HR, Toyota Engineering and Manufacturing of North America.

We are especially pleased how TWI helped us to improve our quality and increase productivity, and how it helped our employees to engage more fully, giving them more satisfaction as we travel farther on this path together.

Chapter 6

Starting Over to Get It Right at Albany International Monofilament Plant

Albany International Corporation is an advanced textile and materials processing company that was founded in 1895. Headquartered in Albany, New York, the company employs approximately 3,500 people worldwide, with manufacturing operations in fourteen countries strategically located to serve global customers. The core business of Albany International is custom-designed and engineered fabrics and process belts, called paper machine clothing (PMC), for the papermaking industry, and Albany International is the world's largest producer. These consumable monofilament, multifilament, and synthetic fiber fabrics are essential for manufacturers to process all grades of paper, from lightweight to heavyweight containerboard.

The Albany International Monofilament Plant (AIMP) was originally founded as the Newton Line Company in Homer, New York. Acquired by Albany International in the late 1970s, AIMP has played a key role in the development of extruded monofilaments (synthetic thread) manufactured using various materials for different applications. The plant offers a wealth of materials expertise to deal with the most unique of customer needs. Products are highly engineered and custom designed to match critical customer requirements. These capabilities enable the company to focus on providing high-value-added products that command above-average prices in the marketplace for product technology, quality, and superior service.

Background

Prior to AIMP's Lean journey and introduction to TWI, AIMP had embraced the Total Quality Control (TQC) philosophy applying Deming's teachings and statistical tools on a regular basis. Process engineers, operators, and supervisors were organized into teams aligned to product application groups (these would be considered value streams today). In response to market requirements, the plant achieved ISO 9001 certification in the early 1990s, and this quality system included a computer-based program for documenting and maintaining operating procedures that was considered world class by third-party auditors. These ISO procedures were used as the basis for their on-the-job training. The system even generated automatic e-mail notifications any time a procedure was created or changed. But, while some formal training or mentoring approaches were used, in many cases this notification was the only training that operators were provided on their job methods. This was especially true for job revisions.

In the 1990s, a significant percentage of operator training was accomplished using a mentoring process. But qualifying an operator in all the required skills procedures using this methodology could take years. Because the operators were organized in product application groups, this meant that even experienced operators sometimes could not fill in if there was a need for them in another application group. Consequently, operator attrition was very painful for the plant. The same could be said for rapid growth.

Due to these shortfalls, the company modified the training process in the early 2000s to shorten the training time and provide for more cross-training. New operators were trained in an intensive operator school, during which time they were relieved of their production responsibilities. This six-week program, developed with guidance from outside adult education specialists, was divided equally between classroom instruction and hands-on training. Upon completion of the program, each operator was assigned a mentor who worked with him or her over a period of several months, refining these newly acquired skills and covering other material as needed. The mentors signed off on all required skills as they were mastered.

During operator school each operator was given a thick binder that contained printed training materials and all the procedures related to his or her position. In total, there were close to four hundred work area procedures and nine hundred documents under ISO 9000 document control, and operators were responsible for following about half of all these documents. What is more, corrective action and continuous improvement ideas, in ever increasing detail, had been incorporated into the procedures for many years. Average work instructions were six pages, and many were in excess of twelve pages. Printing just the first pages of each procedure created a paper stack that was 3 inches high. The end result was that, with so much documentation, there was ample opportunity to deviate from these written procedures.

From Total Quality Assurance (TQA) to Six Sigma

While the plant culture had changed somewhat during a period of expansion in the late 1990s, certain aspects of TQA still remained as the company moved into the new millennium. The need for a flexible workforce and some organizational changes had put an end to teams focused on product application groups, but the need for continuous improvement still existed. A model based less on value streams and more on improvement projects emerged.

The monofilament plant was the first division of Albany International to pursue Six Sigma and, in 2003, generated the company's first certified Black Belt, Kevin Morgan. By 2005, Albany International plants in North and South America as well as Europe were pursuing Six Sigma with guidance from Kevin. Significant numbers of Green Belts were trained, and the corporation established a staff of full-time Six Sigma Black Belts dedicated to quality and continuous improvement. Scott Laundry, an engineer with eleven years in the monofilament plant, took on that role at AIMP and was trained as a Lean Six Sigma Black Belt at the Six Sigma Academy. Scott had been previously trained as a Green Belt and had studied and applied Lean manufacturing while pursuing his graduate degree.

New Players, New Focus

Scott's change in position to be the Six Sigma Black Belt for the Homer monofilament plant coincided with other leadership changes at the corporate and the local level. In April 2005, Scott Curtis became the plant manager, bringing with him over twenty years of manufacturing leadership experience and a passion to improve this major employer in the small town of Homer, New York. With so much manufacturing having left the central New York area in the past several decades, he was determined to make sure this important plant did not suffer the same fate. But it would prove challenging.

The plant had recently experienced significant growth coupled with a change in quality requirements. Scott encountered a plant of 120 people, many of whom were stressed from working mandatory overtime and twelve-hour shifts for months on end to meet production demand. Daily firefighting efforts, due to the increased quality requirements and systems pushed to their limits, had taken a toll on the attitudes of the people, resulting in high turnover rates. The still somewhat lengthy time to fully train new employees increased these frustrations. It seemed to the people at the plant that they might never see the light at the end of the overtime tunnel. Relieving these tensions would become a major focus for Scott.

A second major organizational change was the addition of Tony Wilson. Tony left Lockheed Martin in mid-2005 and joined Albany International as the quality director for the America's Business Corridor. Tony, who has been applying Lean since the 1990s, kicked off a multiplant Lean manufacturing initiative shortly after coming on board. Under his new direction, high-level value stream mapping exercises at each site quickly led to the implementation of targeted rapid improvement events (kaizen events). While AIMP had some success with these improvement events, locking in best practices and sustaining the gains sometimes proved to be elusive. This was especially true when procedural changes needed to be implemented across a wide swath of the workforce. In the meantime, improvement opportunities still abounded and the pressures of the business still remained high.

In 2006, each Albany International plant, including AIMP, participated in an intensive five-day Lean management course, which would become a major milestone in the company's Lean timeline. This event outlined the targets and standards for having various aspects of Lean manufacturing in place. For each plant this signified a significant shift from Six Sigma to Lean Six Sigma (with heavy emphasis on the Lean). A huge amount of content was delivered over the five-day period, including a brief presentation on the five needs of supervisors and skill in instruction (JI) from the TWI program. But few took note. The main focus was on establishing visual controls and leader standard work as described in *Creating a Lean Culture: Tools to Sustain Lean Conversions* by David Mann.* These critical controls allow plant personnel to see performance against expectations and quickly identify problems. When describing the plant's visual

---

* David Mann, *Creating a Lean Culture: Tools to Sustain Lean Conversion,* Productivity Press, New York, 2005.

Lean systems, Scott Curtis refers to these seemingly low-tech boards covered with cards, magnets, and dry erase markers as his daily production reports. AIMP manufacturing can, and has, run for months at a time using these visual controls without computerized business software.

Improved visual planning, along with early efforts at standardizing changeover events and dedicating setup resources, had an immediate positive effect on downtime reduction. The average time for the most common changeover was reduced by 50%. A visual control called the return to green (RTG) board was put into practice and has played a key role in the AIMP Lean journey (see Figure 6.1). The

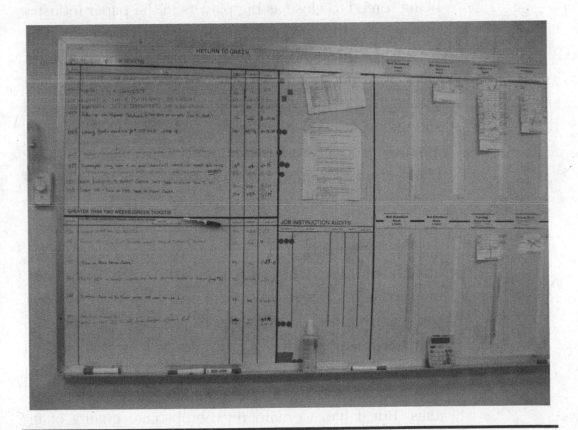

Figure 6.1 **"Return to Green" board. (Reprinted with permission of Albany International Monofilament Plant, Homer, NY.)**

RTG board is used to identify problems, find reasons for missing expectations, and track countermeasures. Pareto charts of RTG items in 2006 began to reveal an unacceptable number of human errors. Eventually, it would become clear that the documentation and training systems at the time were not getting the job done.

Challenging Times

Market conditions in the paper industry during the second half of 2006 were changing rapidly. Many paper mills were being forced to close as big players in the paper industry invested in higher-output machines to reduce cost by increasing throughput. These increased-capacity plants were now meeting the overall global demand for paper. Although the pie was shrinking, Albany International still had opportunities to gain market share by introducing new products and providing technical value to these high-output customers. And, of course, cost, quality, and delivery were still significant factors.

In this tough environment, Lean manufacturing was seen as fundamental to the company's competitive position and to its strategies and financial health. Technology improvements, waste reduction, and a shrinking market had left excess capacity in many plants in the industry. To remain competitive, Albany, as well, was forced to close some of its plants in order to consolidate production volume. These changes put a lot of stress on its systems. But it had to maintain the absolute quality of its production even as it evolved into a "leaner" company.

Creating stability and standardization in one plant alone is challenging. Achieving this across multiple plants when products are being transferred between them and new processes are being introduced is even more so. The company's Lean architects were aware of the role Job Instruction (JI) could play as detailed in *The Toyota Way Fieldbook*, which had recently been published and they were studying. Implementation of the TWI programs had even been included in their overall Lean game plan. Because AIMP was the only monofilament plant in the corporation, supplying this material to other Albany plants that used it to manufacture the final product, standardization across plants was not a direct concern to them, but JI would turn out to be the foundational issue.

TWI's Job Methods program would also play a role as AIMP was charged with helping its sister plants achieve a low-cost position. This meant lowering their own unit costs by increasing utilization and spreading fixed costs over a larger volume of product sales. Ultimately, the increase would mean that the plant would have to produce 17% more tons of monofilament in 2007. AIMP had to become leaner and stay that way.

Early Job Instruction Efforts

Using archival materials obtained from the Internet and *The TWI Workbook*, members of the Albany International Global Training Group created their own set of TWI training materials for the initial ten-hour courses. Mistakenly, though, they viewed the detailed material in *The TWI Workbook* as how to *deliver* TWI training as opposed to how to *use* the method for each program. It was a subtle but important difference that was lost even on very experienced and capable people. They also "enhanced" the ten-hour course training by including standard materials from the Albany International train-the-trainer course, which covered topics like recognizing and adapting to individual learning styles.

Using these materials, eight people at AIMP were trained in JI in early 2007 by an experienced corporate trainer who was AIMP's full-time Black Belt. Five of the trainees were process technicians, those with the highest technical skills in the AIMP hourly workforce. While these technicians were very much in demand for various tasks other than training, the thinking was that the Homer plant would benefit most by utilizing the most experienced people as JI trainers. The training appeared to go well. Over time, though, the AIMP management would come to understand that their initial understanding of JI was lacking and that the training they actually did could not get down to the many subtleties that would allow them to get the full value of using the method.

Looking back at that difficult time, Scott Curtis attributed the problems they experienced getting off on the right foot with TWI to a number of reasons. The internal TWI ten-hour course was put together and delivered by people who did not fully understand what appeared to them to be a simple training method. Although the four-step method of how to instruct and the four points of how to get ready to instruct were clearly presented in the initial training, the learn-by-doing element was only partially effective. Practice training was not properly facilitated due to the inexperience of the trainers who were never taught the importance of people learning from the mistakes of others during these practice sessions. The plant's processes are complex, yet the trainees never learned how to handle special situations (long operations, noisy work areas, getting the knack and feel of a job) in the initial training. In addition, AIMP didn't effectively mentor them through these special cases after the training was over.

The training timetable was also glossed over in the initial training, and the plant struggled with scheduling people for training when there were no replacements so that learners could be taken off their jobs. Perhaps most damaging of all, the mandates they received from upper management for JI breakdown sheets to also serve as reference documents in place of existing procedures slowed the development of the JI breakdowns and made them less effective as training tools. Last but not least, the plant did not have any dedicated resources for JI, and it soon became one more of many initiatives competing for limited resources. Consequently, the ability of JI to stabilize processes and create standardized work never achieved the result that the plant needed.

Problems with Job Instruction Quickly Become Visible

As could be expected, the process technicians, who represented the bulk of the AIMP employees trained in JI, were continually and fully consumed with daily production issues, leaving little time for them to do training and no time for them to create Job Instruction breakdowns (JIBs). The remaining training resources were also stretched thin. Workloads and overtime increased once again as some skilled operators were lost to companies offering more attractive work schedules and job security. Unable to find candidates to fill open positions, the plant continued to struggle with maxed-out capacity and the heavy costs incurred of trying to keep up without the needed manpower. The local training coordinator was fully occupied with just trying to bring new operators on board using the plant's traditional training methods.

Up until the Pareto charts on the RTG board started indicating deficiencies, AIMP's existing training and documentation systems were thought to be already very good. Scott Laundry explains:

> Why fix what wasn't broken? The TWI methods were initially introduced by the quality group as part of the Lean implementation. Although the materials were developed by the Global Training Group, JI was primarily seen as a Lean thing and one small aspect of training in Homer. With much vested in the existing methods, a full plate of training, and JI being pushed from outside, it wasn't surprising that it was not enthusiastically embraced and pursued. A whole book could be written on the missed opportunities for change management.

The bulk of the responsibility for creating job breakdowns, therefore, fell to Scott Laundry, who, in his role as Six Sigma Black Belt for the plant, did see the value of JI. His ability to work on JI, however, was hit and miss given the urgent needs to de-bottleneck the plant and to apply other Lean tools, which took the bulk of his time and effort. Initial breakdowns were primarily focused on locking in new methods developed during kaizen events, and a few were also created to address some chronic quality issues identified by the plant's Lean systems. In addition, JIBs were used for new data acquisition systems in the plant, also a priority.

As mentioned before, the process technicians (PTs) were expected to train operators using these breakdowns. According to Laundry, "Weeks would go by without any training progress. In other cases, training was listed as being completed, but it was obvious that the PTs were not following the JI method to deliver the training. The reason most often given when this was brought to their attention was that 'it took too long.'" This was despite the fact that the JI breakdowns were considered an improvement in documentation over the long book-style procedures within the ISO system. The JIBs were typically 50% shorter, and the critical information was easier to find. AIMP would not understand for many months that even though the JIBs were more concise and followed the format of important steps, key points, and reasons, the initial breakdowns did not support the JI training method because they were nevertheless still restatements of the complete explanation for doing the work.

Frustrations regarding the lack of progress with TWI Job Instruction began to mount as the implementation timeline kept slipping. While there was general agreement that the JIBs were easier to follow than the old book format procedures, there was relatively little pull for JI from the floor. Without the four-step "how to instruct" method effectively applied, a JIB became just another document in the system. Work on JI was sporadic, limited in scope, and limited in effectiveness. Improvements from the kaizen efforts were still not being fully realized.

Deviations from standard work were noted, and some of the same quality issues reoccurred. This caused plant manager Scott Curtis to take a hard look at the status of the JI implementation, where there was certainly great need for improvement. As it turned out, although the plant's problems were already making its personnel sweat, a developing issue would turn up the heat.

The Burning Platform

Throughout this period, production volumes at AIMP continued to increase, requiring a crazy amount of overtime, according to Scott Laundry. Employees were placed under even more stress with the continued loss of experienced operators that left the plant short of people who could perform the complex changeover tasks required to keep the production lines running. A new setup operator position was created to focus more resources on this critical process and to improve changeover performance. However, the plant had no way to quickly train new people in this job who were still uncomfortable with the jobs they were already doing. Lean systems were beginning to reveal that even experienced operators were making plenty of mistakes, contributing to higher costs due to rework, scrap, and lost capacity. Curtis came to the conclusion that they had reached the tipping point and that all of these problems, which had been boiling over in plain view since he had arrived, had reached a stage where they could threaten the viability of the operation. The need for JI was critical, and an entirely different approach was needed—fast.

Turning the Ship Around

At this point, creating job breakdowns was perceived to be the main issue holding them back from success with the JI program. With internal resources stretched, Scott Curtis initiated a search for outside resources by first contacting the training specialist who had helped corporate create the current operator school. The specialist was aware of the TWI JI method but did not know of any outside resources for the specific help with JI that was needed. Scott Laundry had studied *The TWI Workbook* and remembered that the ESCO Turbine Technologies–Syracuse case study in that book involved a local company just 30 miles down the road. They decided to contact "the guys who wrote the book."

Laundry called Robert Wrona at the TWI Institute and found what AIMP needed. Bob introduced Scott to the Central New York Technology Development Organization (CNYTDO), a NIST MEP regional center that provides general business consulting/assistance and delivers programs such as Training Within Industry (TWI), Lean manufacturing assistance, Six Sigma, Lean Six Sigma, Green Belt, and Black Belt training, along with business and manufacturing assessment and small business innovation, in the five-county area in central New York, where the AIMP plant is located.*

Bob was now the executive director of the TWI Institute that was spun off by CNYTDO in 2006 as the nonprofit center for education, training, trainer certification, and networking for the international community of

---

* www.tdo.org.

TWI trainers and practitioners. Hosted by the CNYTDO MEP, the TWI Institute has created a large and rapidly expanding independent network of certified trainers based primarily in the United States, but also with certified trainers in other countries around the globe, who service businesses of all sizes in manufacturing, healthcare, construction, education, and the service industries. Despite the global reach of the TWI Institute, consulting services delivered in central New York are delivered by the regional MEP Center, CNYTDO, whose three Lean Six Sigma project engineers are also TWI Institute-certified trainers delivering TWI locally. This local connection to the MEP would have additional benefits to AIMP beyond just the easy access to TWI experts.

Starting Over with a Mentor

Senior process improvement specialist Cindy Oehmigen from CNYTDO visited the Homer plant to assess their needs and soon became, according to Scott Curtis, an "invaluable resource for AIMP." In addition to providing guidance on their stalled Lean implementation, she also helped to identify sources of funding that might be available to help cover the costs involved in making the changes necessary to keep this manufacturing plant viable in New York State. Of more importance to our story, Cindy focused on helping the Homer plant address their number one priority—to get JI back on track and help them escape from their burning platform.

Cindy immediately contacted Paul Smith, who had been the HR director at ESCO, where he spearheaded the successful implementation of TWI that succeeded in stabilizing processes as part of their strategic plan.* Under Paul's direction, ESCO was able to improve the company's on-time delivery by 80% and reduce training time from two months to two weeks by strategically placing JI in the wax molding department of this producer of precision casting parts. Paul had just recently left his position at ESCO and was now the president of JACM Associates, Inc., a local consulting group that provides consulting services in his primary areas of interest—strategic planning and Training Within Industry.

The CNYTDO plan to speed things up for AIMP was for Paul Smith to create a significant number of JIBs with technical input, as needed, from operators. Instead of just doing that, Scott Curtis decided to dedicate some human resources to work side by side with Paul so that they could learn the process as Paul did his work. Despite the pressures on production, two key people, setup operator Jamie Smith and process technician Susan Rupert, were pulled off their jobs to work with Paul. Jamie had just three years of experience, while Susan was the most senior person on the floor. It was not easy pulling them from production with things so tight. In retrospect, though, Scott's decision to dedicate resources specifically for JI, and the people selected for that task, ended up being critical to successfully jump-starting JI on the shop floor.

When Paul first met Jamie, he knew right away that his job would become easier once "Jamie really got it." Jamie had a strong mechanical aptitude and had developed a good attention to detail while attending diesel mechanic school. Like many young people throughout

---

* For success stories and case studies, see www.twi-institute.org.

the country, Jamie eventually took a job in manufacturing because he got personal satisfaction "making something" while earning a decent wage. Eight years later, the plant in Cortland, New York, where he was working, shut down due to foreign competition, but thanks to his work experience, he was soon employed again at AIMP. Within three years he was promoted to setup operator and also served as a mentor/trainer. From his experience as both an operator and a trainer, he could easily see the benefits of the JI method.

If Jamie's value lay in his enthusiasm and willingness to adopt change, Susan's lay in her vast experience at AIMP and her desire to make sure key points were not missed as the skills and tasks were examined. She could provide the *why* for most of the jobs in the plant. Susan provided a strong sense of confidence that the JIBs would represent the current best way they were doing the jobs when, with so much job simplification going on, there was a concern that important things would be missed.

The plant's training coordinator, Linda Holland, had developed a matrix of training priorities based on surveys of operators and management. The matrix weighed top operator priorities with what management saw as critical to the business. Paul then used this matrix to develop his schedule and assign priorities to his JI team. Having been down this road before at ESCO, Paul knew that his most important task was to mentor Jamie and Sue so they could carry on this work after he left. Paul worked side by side with both Jamie and Sue for roughly two months, during which time Sue came to miss her old job and the pace of working on the factory floor.

Jennifer Pickert, a relatively new operator, was selected to replace Susan, and she rotated in to spend time with Paul before his contract expired. Paul soon learned while working with her that AIMP management had found a good replacement for Susan. While Jennifer did not have Sue's experience with monofilament extrusion, she had Jamie's passion for JI. She also brought with her twelve years of manufacturing experience and eleven years in healthcare, where she was involved with patient care and customer service. Perhaps it was this later experience that made her a natural trainer who quickly developed an easy delivery style using the JI four-step method. The JI effort was now in the capable hands of "the JI team."

Introducing Job Methods (JM) Provides Some Relief

As the Albany International plants began making various degrees of progress with JI, Black Belts were trained and tasked with implementing TWI JM in their respective plants following an established timeline. As with JI, the training materials were developed internal to Albany International, this time by Master Black Belts who also used *The TWI Workbook* as their primary guide. Scott Laundry noted, "The course was close to the original in content if not completely consistent in delivery. The examples and steps were the same, but group exercises were conducted instead of having each trainee create a JM breakdown on a job from his or her workplace as required in the original TWI JM course. Having fewer breakdowns limited the learning, but some good results did come from the training. JM has many similarities with classical tools used by Black Belts, like process mapping, and individual Black Belts felt comfortable coaching and mentoring the course."

The initial AI JM training was well received in Homer, but resources once again became an issue. Adoption of a new computer business system was looming, and area managers had to devote increasing amounts of time to support the upcoming change. JM was largely seen as important but not urgent. In manufacturing, the struggle to reinvigorate JI was also competing for priority. Paul Smith was just getting the JI train back on the tracks and Scott Curtis did not want to run the risk of derailing it again by pushing JM too hard.

In one department, though, JM became a very high priority. The warehouses were, at that moment, the plant's bottleneck. Warehouse manager Ben Stuttard had been trained in JM and saw the potential for his department when a JM proposal in the ten-hour course had eliminated three hours a week of waste in the warehouse with no expenditure. The number of people involved with material handling was relatively small, and many of the potential improvements could be effectively implemented without extensive training.

JM proposals in the warehouse, backed by documented savings, were effective in overturning sometimes onerous practices that were dictated, oftentimes, by outside departments. In one example, established practices required labeling, lifting, and transportation of individual 50-pound bags of plastic pellets to satisfy an inventory replenishment system. Application of JM resulted in a simple *kanban* system, with the original pallet of fifty-six bags as the container size. This improvement alone reduced handwork and material handling by 98% without increasing inventory. Another JM breakdown generated an 80% reduction in order picking time by rearranging the picking activity to better correlate with the creation of the final picking list. Average improvements for other tasks were around 50% and eliminated three hours of non-value-added work per week.

As JM freed up time and overcame chronic frustrations with inefficiencies, an interesting thing happened. Warehouse personnel would ask, sometimes daily, if JM could be applied to specific tasks in order to make them better. In other words, JM was proving to be an effective tool that was developing grassroots pull from the people on the floor. Eventually the new methods, which often included classical Lean tools such as 5S and kanban resulted in shifting the plant's constraint from the warehouse back to manufacturing, where they were still in the early stages of establishing effective TWI Job Instruction. JM for the shop floor would have to wait until competence was established in JI, just as it was done at ESCO.

It was now clear that the right people were involved, and the improvements coming from the JM activity provided just the measure of relief the JI team needed in order to have time to grow and make progress.

Initial Lessons Learned from the CNYTDO Mentoring Project

It was clear to Paul Smith where Albany International had missed the mark when they started to implement the JI program. In the past, training had been inexorably linked to verbosely written documentation. This, and a corporate mandate that JIBs would replace, not supplement, existing procedures, had resulted in initial job breakdowns that were too long and too complex for trainers to use with the JI four-step method. For example, one trainer covered seventeen important steps in a first-generation JI breakdown of a complex computer interface setup that took approximately one and a half hours to perform. At that rate, it would have taken ten hours to train the whole job, and one can only guess how many times the trainer deviated from the JI four-step method. This breakdown was more a reference document than a tool for training, and that is a mistake many companies make when they think they understand how to break down jobs without getting a proper understanding of how the JIB is used. (See Chapter 4 for a complete explanation of the difference between SOPs and JIBs.)

Paul also trained Jamie, Susan, and Jennifer to handle the special instruction situations that come up, just like this example of a long, complex operation. He taught them how to break the job down into smaller tasks, or teachable units. In so doing, they learned that the only purpose for the JI breakdown is to help the instructor train a person to do a job. If the trainer is unable to train the person using the breakdown, then the problem is with the breakdown, not with the person being trained. Paul worked with the team to break down this job into three segments, each having a separate JIB. Information not essential to training the operators was moved to a separate reference document. After simplifying the process in this way, each segment actually required only fifteen minutes of total training time.

Setup Operator Training

Weeks were invested in creating new job breakdowns before Jamie began training the setup operators on critical tasks. The first job he trained was the disassembly and cleaning of a gravimetric feeder system. When getting everything ready to train, Jamie disassembled the feeder in order to teach how to assemble it only to find that it had not been assembled correctly the first time. If someone operated the system as he had found it assembled, the feeder could have been the source for processing problems and defects. While training other setup operators, Jamie noted similar errors in the assembly of the feeder when preparing to train. Upon investigation, he learned that each assembly error found had been made by someone not yet trained using the TWI JI method.

As the training progressed and he did regular follow-ups, Jamie noticed fewer setup errors as more setup operators were trained. Errors essentially ceased after all setup operators had been trained using what were now well-written JIBs that reflected best practices. Feedback from experienced operators after training let Jamie know they were on the right track, with operators making comments like, "I learned something new" and "I never knew why we did it that way." The high number of errors encountered while training on this first critical task was an early indicator of the impact Job Instruction would have at AIMP. Human errors by operators were eventually reduced by 70% overall when personnel were trained using the JI method as shown in Figure 6.2.

As Jamie started to train significant numbers of operators, Paul Smith began tapering off his coaching and

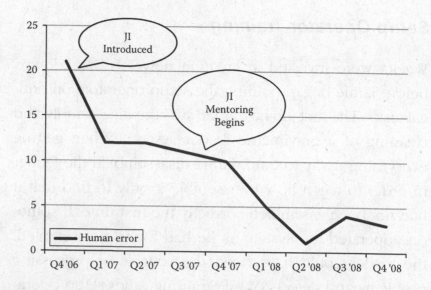

Figure 6.2   Human errors by quarter. (Reprinted with permission of Albany International Monofilament Plant, Homer, NY.)

letting the team take over on their own. Jamie quickly experienced problems coordinating training sessions in the 24/7 continuous manufacturing environment. So he used his training on timetables to convert the training priorities provided by Linda into an easy-to-follow visual. This was communicated to first-line supervisors. For any given task, less experienced operators were trained first based on the assumption that inexperienced people would be the source of most errors. This led to some experienced operators asking when *they* were going to be trained in the new method. These operators wanted to do their jobs well and weren't always 100% comfortable doing the tasks. They didn't like being at the back of the line. These were all good signs that JI was having a positive impact on the attitudes of people and that it was being accepted into the culture of the plant.

Perhaps the most telling sign that JI was taking hold was the sign that someone had pinned up outside the trainer's office, which read: "Home of the JIBLETS." This good-natured fun was a sure indication that they were getting people's attention and a high level of acceptance. It was also a sign that employee stress had begun to wane.

Making JI Training Stick: The Auditing Process

As JI began to take hold and spread, the importance of what Lean practitioners might call job checking became evident. AIMP called these checks JIB audits and used the form shown in Figure 6.3. An audit involves having

JIB Audit Sheet	
Person audited:	Date of audit:
JIB reference code:	Rev. (Ver.):
Description (JIB title):	
Comments from auditor:	
Auditor:	Additional training needed? No/yes Method change needed? None/immediate/future rev.
Feedback from employee for suggested changes:	
Problem:	New method proposed:
Results if new method adopted:	

Figure 6.3 JIB audit sheet. (Reprinted with permission of Albany International Monofilament Plant, Homer, NY.)

a qualified observer watch another worker perform a job while utilizing the JIB to make sure the worker is performing all the important steps of the job and hasn't missed any key points. A successful audit provides an opportunity for the observer to thank the individual for a job well done, and if a deviation is noted, the chance to discuss with the worker why it happened and identify the root cause. Retraining by a JI trainer may be required, and if multiple audits revealed similar problems, the next step would be to question the effectiveness of the JIB. Maybe the reasons for the key points were not clear or compelling enough. If it is an isolated incident, a trainer might have forgotten to get the person interested in the job, which is a part of Step 1, Prepare the Worker. At AIMP, the mantra "If the worker hasn't learned, the instructor hasn't taught" is taken very seriously to maintain a positive culture and environment.

Observers also learn that there can be many other reasons why people deviate from standard work. The worker might be doing a best practice that was overlooked when teams made a new breakdown, even as they tried to build consensus around the method. It is rare to be able to observe one person and have the breakdown represent *the* current best practice. Performing the JIB audits gives trainers the opportunity to note possible best practices that can be incorporated into the standard work by observing a number of operators doing the same job. Without such observation, feedback, and retraining with the latest JIB, the impact of determining new best practices might remain isolated to that one individual.

Shift supervisors (called coaches at AIMP) were not forgotten in the JI implementation process. While the training has been delegated to trainers who lead the work of others by training them to do the jobs, the traditional supervisors perform the JIB audits as part of their leader standard work. Audits by shift leadership primarily serve to reinforce their commitment to standard work and to provide opportunities for leaders to thank or coach people as needed. Sometimes, though, despite good training, people stray from the path. This is one of the many times that come up where trainers rely on the Job Relations four-step method to gather all the facts to determine if the person has the ability to perform that job or if the person might in fact be better suited to perform another job within the company in order to "make the best use of each person's ability."

Sharing Best Practices and Lessons Learned

During the period Paul was working with the Homer plant, training coordinators from Albany International plants in North America were invited to attend a Job Instruction workshop. This workshop was facilitated by CNYTDO and structured so that AIMP could share best practices and lessons learned with other plants. Not unexpectedly, many of the pitfalls initially experienced by the Homer plant had been repeated in other sites. An overview of the JI method was given to emphasize the importance of one-on-one training and adhering strictly to JI's four steps for instruction. A demonstration of proper JI training using a breakdown was also given. Existing breakdowns from each plant were displayed and were evaluated with the same process used for breaking down demonstration jobs in the ten-hour course. As the workshop progressed, some training coordinators began to critique and improve their own breakdowns as they were presented. The training coordinators left AIMP with a significantly improved understanding of JI and the tools to improve the training in their own plants.

This sharing was just the beginning. In 2008, the Homer plant opened itself up to visitors from other companies and began to share Lean practices and to relate their experiences with TWI to a significant number of tour groups coordinated by the CNYTDO. Even though the visitors were coming to AIMP to see what they had achieved, it proved an excellent opportunity for AIMP people to learn from and benchmark with others. The sharing of TWI insights also extended to Albany International plants in different divisions and on different continents. Within a year the plant had progressed from having to be mentored by TWI experts to becoming a mentor for others in the TWI process.

Shifting Gears

During the second quarter of 2008, hiring, training, and efficiency improvements finally helped the plant catch up to production demand, while the stress due to the increased volume at last became manageable. The plant had survived the burning platform it faced in 2007. AIMP had realized significant results since starting its Lean journey and introducing the TWI methods. In addition to the 70% reduction in errors mentioned above, net profit at AIMP rose more than 250% from 2006 to 2008, as shown in Figure 6.4.

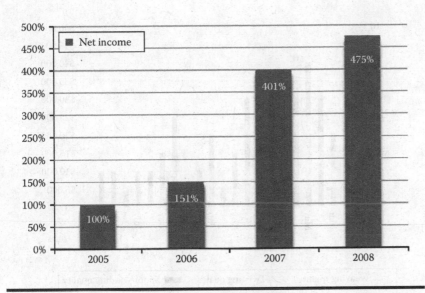

Figure 6.4   Net income 2005–2008. (Reprinted with permission of Albany International Monofilament Plant, Homer, NY.)

Quality delivered to AIMP customers also improved. Customer complaints and irritations were reduced to historical lows, and returns and allowances approached a new low point of just 0.01% of sales. While all of the improvements in Figure 6.5 cannot be attributed solely to TWI, TWI played a significant role as a strategic part of their Lean Six Sigma strategy.

After the highest-priority setup tasks had been trained, the focus of the JI implementation then shifted from setup operator tasks to a wider scope of operations. AIMP employees recognized that other jobs beyond the setup tasks could benefit from JI, and this caused an increase in demand for TWI training. A new expectation had emerged. New equipment in the plant wasn't put into service without a job breakdown and training to ensure correct operation and safety. Anyone in the plant trying to implement a new procedure, change an existing one, or lock in an improvement realized that

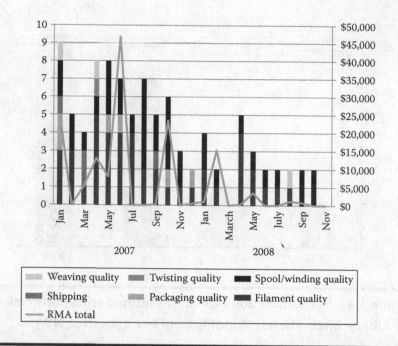

Figure 6.5 **Complaints and irritations by issue. (Reprinted with permission of Albany International Monofilament Plant, Homer, NY.)**

the most effective path to getting that done well was to involve the JI trainers. What was at first viewed as a temporary assignment for Jamie Smith and Jennifer Pickert to complete the JIB project had morphed into permanent assignments for both of them as dedicated resources to support the TWI training. Fortunately for AIMP, both Jennifer and Jamie were more than interested in continuing in these roles.

Expanding the TWI JI Infrastructure

Scott Curtis decided that in addition to having dedicated JI trainers, AIMP needed a certified internal resource to allow the plant to broaden the scope of JI beyond the existing core group. This time the plant would follow the proven TWI JI delivery method. During Paul's mentoring period, the initial group of JI trainers were all able to attend a ten-hour JI course delivered by Glen Chwala, a TWI Institute-certified trainer from CNYTDO. The ten-hour course, which provides the basic tools needed to break down jobs and train them using the JI four-step method, is also a prerequisite to attend the forty-hour train-the-trainer program for Job Instruction, where trainers learn how to deliver the ten-hour course themselves. Jamie was selected to become a certified JI trainer by attending this training conducted by a TWI Institute master trainer in the third quarter of 2008.

Jamie, who had started out as an extrusion operator and progressed to setup operator, now had a new career as TWI Institute-certified JI company trainer for Albany International and had this to say about his new role:

> I was excited to be chosen for the job. It was a great opportunity to help production learn and to understand their jobs. I developed a passion for TWI, and I knew that it was the right career for me. As an extrusion operator my ability to help other plants improve was limited. Now I have a universal skill that allows me to help the entire company.

Jamie used his advanced training to train and certify more JI trainers at AIMP as the pace of training increased several times over. Currently, the JI trainers in the plant deliver about three thousand one-on-one JI sessions per year, or about one per week per hourly employee. This is a reflection of their enthusiasm, their competence, and the demand they have helped generate for good training. Jamie has received many comments from operators in the plant regarding JI since training began. Some of these quotes are provided below:

- "I learned something new on a blender system after using it for two years prior to the training."
- "Works out very well now that new trainees are fully trained in one week rather than three, like before."
- "The training was really down to earth. With the repetition it was easy and fast to learn the job."
- "The JI training is a very effective way to get employees to do a job the correct way and to keep a standard way of doing it."
- "Writing breakdowns is a great tool to look at a job and find waste and remove it."
- "I have operators asking me almost every day when I am going to train them. It is a really positive thing considering I thought it would be the opposite, with some push back and attitudes."

Not every new program receives these kinds of positive comments. It is a good indication that JI at AIMP will not be just another flavor of the month that withers for lack of visible benefit.

New Challenges

As JI was spreading, the adoption of a new business system had created a whole new set of challenges that demanded managers' time and focus. People were first focused on looming deadlines, and then on working out the bugs in the new system. This multitude of efforts also severely limited the Job Methods activities they had begun in the warehousing area with some success. Ultimately, the global economic crisis that began in 2008, where Albany International also experienced a significant downturn, provided both the opportunity and the necessity to renew Job Methods at AIMP.

AIMP took this opportunity to have two members of each shift trained in Job Methods so that in the future improvements would not be solely dependent upon area managers. Each person who went through the training was given a chance to present his or her JM proposals to the AIMP leadership team after completion. As with JI, Job Methods generated positive comments, such as:

- "It's amazing that you can do these jobs every day and not see the waste until you break it down using JM."
- "It didn't seem like there was a lot of waste, but when you add it up, it's huge."

Their positive comments are easy to understand when considering that applying the method to the ten simple tasks selected by the ten-hour course participants alone identified more than $100,000 per year worth of savings.

In order to support the larger Job Methods effort, Jennifer Pickert was asked to expand her role beyond Job Instruction to include working with the shift JM teams, coordinating and assisting them with their improvements. She plays a key role in this effort, oftentimes serving as a sounding board for people working out their ideas. She also helps ensure that the new methods are actually applied and reap their promised benefits, by helping team members prepare their proposals and then making sure they lock in the improvements by teaching them properly to operators using the JI method.

Talent Management

Selecting the right people for the right job is one of the critical steps that leaders sometimes fail to give proper attention to. This is even more true when it comes to implementing a successful TWI rollout, as can been seen in all the case studies in this book. Susan was the most senior person on the shop floor and was respected by her coworkers. Most companies would normally view her as being "too valuable" to move out of production to create JIBs, a task that "anyone can do." The truth is that it is because of her valued status that she was able to correctly get the breakdown process off the ground, a task that not just anyone can do. For Jamie and Jennifer as well, although both had only three years of experience working at AIMP, they had rich backgrounds and were in key positions to help meet production goals.

One of the foundations for good relations in JR is to make the best use of each person's ability, and we do that when we look for abilities not now being used and never stand in a person's way. These principles worked well for AIMP. Jamie stepped beyond his initial role as a setup operator and is now the TWI-certified Job Instruction company trainer for AIMP and an important resource for other Albany International plants that need help with JI. As one of the most knowledgeable operators, Susan's contributions to creating JIBs were extremely important, yet Scott Curtis honored her request to go back to her old job when it was clear her passion remained with being a process technician. Jennifer, in turn, has done such a good job of developing JIBs and using them to train people that her role and responsibilities, like Jamie's, have greatly expanded. Her passion for continuous improvement has led her to take on key roles in program follow-through in both JI and JM.

Closing Remarks from Plant Manager Scott Curtis

As the global downturn hit bottom, Albany International was dealing with many tough decisions, one of those being how to further reduce raw material costs. Faced with several suppliers running at less than full capacities and looking for opportunities to reduce costs through increased volume, we had to make some tough choices. One of those choices hinged on our ability to quickly absorb more volume without an interruption in service levels or quality. Based on the gains experienced over the last three years, in large part due to the TWI and Lean methodologies implemented, AIMP was fortunate enough to gain the market share to become the key monofilament supplier to the company. This decision did, however, result in the plant closure of another supplier. This could have been a different outcome for AIMP had we not experienced the improvements we had over the last three years.

Moving forward, AIMP is in a much better position to react. The plant now has the training infrastructure and an improved ability to quickly respond to a rapidly changing world.

III

TWI AND CULTURE

Chapter 7

Building a Culture of Competence: The Human Element of Each TWI Program

One of the most emphasized aspects of any successful Lean implementation is the importance placed on people in making it all work—what is known in the Toyota Production System (TPS) as respect for people. Without the cooperation and input of people, in particular those people who do the work, we will not get the buy-in necessary to see that needed changes in fact take place and are sustained over time. It is easy to become enamored with technology and tools—aspects of the Lean transformation that we can actually touch, see, and do—while overlooking or neglecting the less obvious, but more demanding aspect of the people involved. David Mann, author of *Creating a Lean Culture*, believes that in a truly successful Lean conversion, 20% or less of the actual effort goes into physical changes in the production process in terms of things like new layout to establish better flow of materials and people, pull systems to synchronize production, visual tools that reveal abnormalities in work, and so on. In fact, Mann feels that this is "the easiest fifth of the process!"[*] What remains is how we deal with people in managing daily routines and the communication and problem-solving processes. These things make up the management system so crucial for the long-term success of any Lean implementation.

Organizations are, by definition, "group(s) of people identified by a shared interest or purpose."[†] And we know that when we ignore the human element to our organizations, trouble is certain to ensue, preventing us from achieving that shared purpose. When we nurture those human relations, on the other hand, we open up the opportunity for motivating people to achieve because they identify with the shared interests of the group. John Shook explained what this process looks like at Toyota in his book *Managing to Learn*:

> My own epiphany came when my boss told me, "Never tell your staff exactly what to do. Whenever you do that, you take responsibility away from them." His comments revealed how Toyota operates not as an "authority-based" but a "responsibility-based" organization.[‡]

Because Japanese companies like Toyota have been so successful, both inside and outside of Japan, at building strong relations with their workforce, it is commonly believed that this philosophy of engaging people in how

---

[*] David Mann, *Creating a Lean Culture: Tools to Sustain Lean Conversion,* Productivity Press, New York, 2005, 4.

[†] *Encarta Dictionary: English* (North America), http://encarta.msn.com.

[‡] John Shook, Managing to Learn: Using the A3 Management Process to Solve Problems, Gain Agreement, Mentor, and Lead, The Lean Enterprise Institute, Cambridge, MA, 2008, 2.

to do the work came to us from Japan. That is not so. When interest in Japanese management practices began to stir here in the United States in the early 1980s, the notion of asking operators for their ideas and opinions on work methods, something the Japanese were doing to great success with their kaizen approach, was quite revolutionary. In those days, it was a common belief and practice that, at least in mainstream American companies, the white-collar people would do all the thinking and the blue-collar people would do all the working. Strangely, though, when Patrick got to Japan in late 1980, just after graduating from Drexel University, the Japanese managers he met there were dumbfounded at all the attention being given to their way of management. "Why is that?" they asked. "*You* taught us everything we know."

The contribution W. Edwards Deming made to Japanese manufacturing in the postwar era is well known, and more recently, we have come to realize TWI's contribution to the Japanese manufacturing resurgence following World War II. In his groundbreaking research reintroducing TWI into the American arena, Dr. Alan Robinson reports that, in an interview from August 1951, the "concept of humanism in industry" was "one of the most appreciated ideas transmitted into Japan by TWI."* The notion that good management included a respect for individuals was not, at that time, a part of the Japanese style. Robinson claims that, in addition to developing an appreciation for a more rational approach to management, TWI was able to teach the Japanese that "good

---

* Alan G. Robinson and Dean M. Schroeder, Training, Continuous Improvement, and Human Relations: The U.S. TWI Programs and the Japanese Management Style, *California Management Review*, Vol. 35, Winter 1993.

human relations are *good business practice*, a message that is given credit for helping break up the tradition of autocratic management prevalent in Japan before and during the war."*

The Japanese, then, learned from the Americans that it was important to emphasize the role and participation of people, including operators, in getting the work done well. In the Job Methods module of TWI, there is a large sample demonstration done in Session 1 to illustrate and introduce the four-step method to JM. One of the critical components of this presentation is to highlight how the supervisor worked with one of the operators in applying the four-step method to find improvements to the process of "making and packing of radio shields." We still use this demonstration today in its original form, with the only changes being the name of the product (microwave shield instead of radio shield since radios have become quite smaller), the name of the operator (Bob Burns instead of Jim Jones because of the tragedy associated with the original name), and the name of the supervisor (Anne Adams instead of Bill Brown to provide more gender diversity, which was lacking in the 1940s' courses). The original manual dialogue specifically points out which ideas were given by Jim Jones and goes on to highlight that contribution: "Remember how Bill Brown 'worked with' one of his operators? Operators have good ideas too; often just as many as we have—sometimes more!" It goes on to further emphasize the importance of getting buy-in—new-age terminology for an age-old concept: "When he [the operator] helps work

---

* Ibid.

out an idea he gets real satisfaction. An interested and satisfied worker is just as important as the idea itself."

It's clear that the practice of asking workers their ideas and opinions is one that we, in the United States, had at least as far back as the 1940s, when the TWI programs were developed. Knowing that, it is now easy to understand why the Japanese were so puzzled at how enamored Americans were with Japanese management's respect for people—a practice they learned from the Americans.

The Human Element

When doing training at manufacturing plants, we usually find supervisors are hesitant to approach what they call the touchy-feely realm of human interaction. After all, they say, that's why they got into manufacturing— because they didn't like the peculiarities and difficulties of dealing with people and had much more proclivity to working with mechanical things, things that didn't talk back or "have a bad day." These supervisors see themselves as masters of the physical universe where problems can be solved rationally and where, with just the right amount of organization and discipline, everything will work just the way it is supposed to. When people problems do come up, they brush them aside saying, "Leave your personal issues at home, we're here to get the work done."

Of course, good supervisors have learned from hard experience that this kind of attitude toward people simply doesn't work and things never go the way they are supposed to. In fact, one of the most frequent remarks we get back after giving the Job Relations module of TWI, where we learn to solve and prevent these very people problems, is: "I wish you would have taught me this ten years ago. It would have saved me a lot of headaches."

Here, again, we see the wisdom of the original TWI training courses. The Job Relations training manual from the 1940s recognized this need to treat people with respect because each person is a unique and capable human being. It is only through the "loyalty and cooperation of people, in addition to what machines can accomplish," the manual stated, that supervisors can achieve the output and quality they are charged with delivering. The manual specifically points out that people must be treated differently than the machines they operate, a point that must still be taught today, just as it was in the 1940s:

> When a machine is installed in a department, a handbook comes with it—or there may be a mechanic specially qualified in how that particular piece of machinery works, and directions on how to keep it in good operating condition, or what to do when it breaks down.
>
> Supervisors get new people all the time, but handbooks don't come with them.
>
> How are you going to keep those new persons in top form? What will you do if they fail?

The entire Job Relations course teaches a method for leading people effectively, and here let's consider the results of learning and using the method and the effect it has on developing true respect for people, that vital element of Lean.

There is a good reason why untrained and inexperienced supervisors avoid getting to know their people as individuals. They hesitate because they are afraid employees will demand something they cannot give: higher wages, better working conditions, a smaller workload. They are also afraid that if they get too close to their people, that will prevent them from exercising real authority and discipline when needed. Getting close seems like they are "making friends," and employees may then demand "special treatment," or at least it will look that way to others. At the very least, they fear, if supervisors become too familiar with their people, they will lose their aura of superiority and rank. Without that positional power of infallibility, they may find themselves having to be truthful with these people, whom they have gotten close to, including admitting fault when they are wrong.

When we learn and practice the JR skill, though, and become good leaders, these fears will prove to be unfounded. In fact, what we have found by doing countless JR practices during our training sessions is that supervisors shouldn't assume they know what employees want—more money, better conditions, etc. These assumptions are almost always wrong because we view these problem employees from our own perspective, from our own personal histories. If, for example, I had a strong-willed parent or teacher in my childhood who "whipped me into shape" when I was wavering, I will assume that everyone else needs a similar experience when they run into difficulties, and I will try to be that same disciplinarian for others that worked for me. But we don't, in fact, understand their true motives. These people are different and have not had our same life experiences. What worked for us will not necessarily work for them.

What is needed is for us to treat people as individuals and to take the time to understand them as unique human beings. This is the real meaning of respect for people. JR teaches very specifically and emphatically that in order to get the facts as the first step to effectively solving a people problem, we need to get the opinions and feelings of the persons involved. In fact, it goes on to give tips on getting those opinions and feelings, advice such as "Don't argue," "Don't interrupt," and "Don't do all the talking yourself." In order to lead people well, we cannot afford to ignore their personal issues. Just the opposite, we need to hear them out in order to "be sure you have the whole story."

This practice goes against our fears of intimacy with people but, when done purposefully, can help lead us to finding the root cause of their problem, which in turn will allow us to find the solution that solves the *real* problem—not the problem we assumed it to be. Once we find out what the real problem is, the solution, it turns out, will not cost a lot of money—though it may cost the supervisor some amount of pride. You may have to say, "I'm sorry." In the end, the supervisor has what his or her people want, and it is usually not difficult to give. The real difficulty is finding out what those needs are. And that's where the JR method helps. By effectively resolving these issues, we build the trust and respect of our people, and this far overshadows any concerns over losing positional authority.

One example of a problem a supervisor brought into a JR class illustrates how the JR method helps develop a real respect for people. A new supervisor had a person under his charge who had been a quiet but diligent worker at the company for many years. The company had never had any problems to speak of with this person. Then, around 2004, the worker became argumentative and difficult to work with for everyone in the shop. The young supervisor warned him about his behavior and how it was adversely affecting both the work environment and the output. But the worker was angry at this admonishment and his behavior got even worse. The supervisor tried again to talk with the person, imploring him to set a better example for the others, but it did no good.

Cases such as these follow a typical pattern. As the person's behavior continues to deteriorate, he or she ultimately commits some "unforgiveable sin," such as refusing an order or threatening another employee, and as a result, he or she is eventually, after many warnings, terminated. When this happens, supervisors justify the action taken by saying that, in fact, these employees "fired themselves" with their unacceptable actions. Unfortunately, and this is another common trait to these sad stories, the fired person was almost always a good worker, maybe one of the best in the department, and a replacement will now have to be found and trained—a difficult and expensive process that does not always bring you a better or even equal person to the one let go.

In this particular case, though, the young supervisor applied the JR technique of getting opinions and feelings and made a big effort, in spite of the hostility received from the disgruntled worker, to find out what was really the cause of the bad behavior. It turns out the man was a Viet Nam War veteran and was angry upon seeing the hero's welcome soldiers returning from the current war in Iraq were receiving. "When we got back from Nam, we didn't get no parade," he said bitterly. He explained how it was difficult to talk to anyone about those old wounds, especially to a young person like the supervisor, who had not lived through that era.

Now that the problem was better understood, the supervisor could think of more possible actions that could be taken that would get to the root cause. This part of the JR method is the most critical, for it recognizes that people are different—they are all individuals—and that no one action can apply to everyone because the cause of their behavior will never be the same. And without acting on the true cause of the behavior, we'll never come to a satisfactory solution. Veteran's Day came up shortly thereafter, and it was a holiday for the company. On that day, the supervisor called the man at home and, while saying he didn't want to bother him on his day off, told him he was thinking about him and thanked him for the service he gave to his country.

The action this supervisor took didn't cost him, or his company, a single dime. But, at the same time, it wasn't an easy thing for this young man to pick up the phone and express his sentiments after all the hard feelings that had gone down between the two of them. He had to swallow some pride and admit to misunderstanding initially the employee's predicament. His action was effective, though, as the person's behavior returned to normal and no adverse actions were taken against the employee that would have been regretted later.

This example shows the power and simple genius of the TWI JR method, and it demonstrates how the TWI methodologies are engrained with that human element that helps us deliver on respect for people. While companies are certainly not in the business of caring for people's personal needs, and the final checkpoint of the JR method is "Did your action help production?", we learn that when we do right by our people, good results follow.

Motivating the Workforce

It's common to hear the complaint that workers have a "bad attitude" or "all they care about is money." Every generation seems to complain about the next one when it comes to work ethic, as if they were the first ones to know and value the benefits of hard work and dedication. And when we see this poor attitude toward work, we instinctively think that there is nothing we can do about it. Yet these same people with the poor attitude most likely entered the company full of zeal and bright dreams. When they had their job interviews, they may have even promised to be the best worker the company ever had, and they meant it. Somewhere along the line they lost that motivation.

Stephen Covey, in his monumental work *The Seven Habits of Highly Effective People*, taught us that for those closest to us, the ones we see and interact with every day, we need to give those relationships close and regular attention. It may seem strange, he said, but for people we haven't seen or talked to for an extended period of time, we can pick up those relationships right where we left off with them as if not a day had passed by. Yet for our loved ones and those we see each day, we make the mistake of thinking that exactly because we see them on a regular basis, we do not need to attend to those relationships. The opposite is true. These are the relationships that vitally need our regular and continued attention.*

The Job Relations element of TWI covers this aspect of human nature thoroughly by presenting four foundations for good relations that, when practiced on a *regular* basis, help prevent problems from happening. In fact, from experience we have seen how most any human relations problem, when we search for the root cause of that problem, has its genesis in some violation of these foundations. For example, we know from a multitude of employee surveys over many years that the number one reason why employees quit their jobs is because they feel they are not appreciated, a violation of the foundation "Give credit when due." Even when employees do fail, they still feel they have been wronged because they were never given a chance to correct their work, a violation of the foundation "let each worker know how he or she is doing." Then there is the often heard remark, "It's not *what* they did, but *how* they did it." In other words, it's not that employees are mad at what was done by management—though they certainly didn't like it— but it was how it was done that created the lasting hurt and anger. The foundation "tell people in advance about

---

* Stephen R. Covey, The 7 Habits of Highly Effective People, Simon and Shuster, New York, 1989.

changes that will affect them" helps to prepare the way for change so that people have a chance to have their say and don't feel they were "stabbed in the back." Finally, when we follow the foundation "make best use of each person's ability," we prevent problems where people feel they have been ignored or passed over.

By taking care of these basic human needs and providing continual attention to people, we keep our relationships with people strong and prevent good attitudes from going sour. It's not enough to complain about people's "bad attitude" without first considering our own responsibility for allowing that bad attitude to infest and fester. This responsibility is at the core of good leadership.

The Job Instruction component of TWI also covers this vital element of attention to people. Step 1 of the JI method is to Prepare the Worker, and here we take a few moments to find out about the learners and what they know about the new job to be learned, and, most importantly, get them interested in that job so that they want to learn it and continue to do it carefully and con-scientiously from here on out. When employees are new and know very little about the work of the company, we usually, though not always, cover many of these points in some type of orientation or "on-boarding" program. But these are general, overarching themes, and they are typically only expressed once, when the person first walks into the company. With JI, supervisors must take the time to prepare all of our people, even the veterans, for learning jobs as we teach them, one by one.

For example, when a person learns a new job he or she has never done before, this person may have had similar jobs or duties in the past that he or she can relate to that will help to learn this new one. The trainer should find this out right up front—that's why it's in Step 1. But JI goes further. Even if the person has never done a job like this one, we try to find out if there are any other things he or she does in life, outside of work, like a hobby or a sport, that use similar skills or experiences. Many people do work at home or in the garden or with their car where they use similar tools and do familiar operations. A typical case is where the trainer asks if the learner enjoyed building plastic model kits as a child, because the assembly work he or she will be learning is "just like that," using similar skills.

In the same way, the best way to get a person interested in a job is to show why the job is important and what happens if it is not done correctly—let the person see the bigger picture of how his or her small, but important, contribution affects the final product or service. Workers may think of themselves as insignificant cogs in the bigger machine, and they'll never know their own role unless someone tells them. Supervisors, after learning to apply this aspect of good job instruction, are continually astounded at how little their operators understand about the significance of the jobs they perform each day. They assumed, incorrectly, that the operators know these things. At a shipyard, one operator was surprised to hear that the measurements he made on cutting pieces of metal actually could lead to large sections of the ship not fitting together. The operator didn't know where the plates would be located on the ship.

Even when learners don't have the experience to understand how their new job will fit into the larger work processes, you can still relate the job to things they may do in life. When one trainer was teaching a person how to a prepare a spare parts package that went with a maintenance kit, he asked the learner if his wife had ever sent him to the grocery store to pick up a particular ingredient she needed for a recipe she was working on. The learner said she had. Then the trainer asked if he had ever brought home the wrong thing—the wrong size or amount, the wrong brand, the wrong flavor, etc. Nodding his head with a wry smile, the learner said he had. "How was that?" the trainer asked him. "Not too good!" was the reply. "Well, that's what happens when you don't check the labels carefully and you put the wrong parts into this package. The installers get really mad." By relating the learner's personal experience with customer satisfaction, its importance became more than just an abstract idea.

One of the most effective elements of the Job Instruction method (many feel it is the *most* important part of JI) is to teach the reasons why we do the job in a particular way. In other words, when we point out key points—critical elements that "make or break" the job, injure the worker, or make the work easier to do—we must tell the learner the reasons why we have to follow these key points. For example, if the key point when tightening a nut is "hand tight, then half turn with wrench," by telling the learner that the reason for this key point is because "it will be difficult to remove if overtightened," we engage the person's brain and have the learner think conscientiously about what he or she is doing. And this reason needs to be the actual reason, not just a platitude like, "It's the standard" or "That's the way it's always been done."

As we said earlier, people are not machines and should not be treated as such. Yet supervisors, consciously or not, seem to expect people to do their jobs mindlessly without knowing why they should follow a specific procedure. A machine doesn't need a reason for performing an operation, but humans, who possess brains, are influenced by their understanding of a job as much as they are by the physical act. This respect for human ability, their capacity to think and to understand, will further motivate the person to remember and follow the standard work since people are not inspired to do things that have no meaning. After all, why would they do something "for no reason?" If we want people to care about what they're doing, then they have to know why it is important, and then they'll remember to do it that way each time.

The plant manager of a German multinational auto parts manufacturer gave the example of an operator on a chip line who didn't want to follow the standard work because there was no obvious reason to do it that way. He was a bright young operator and questioned the method because it seemed, from his vantage point, to be wasteful. The situation was that the reason for doing the key point came up in a later operation that no one had ever told him about. The plant manager walked the operator 100 feet down the line and showed him why his operation had to be done that way, and the resistance to the standard work disappeared.

In all these ways, the TWI methods focus on respecting the people who do the work and dealing with them as human beings. In the hustle and bustle of a usual day, it is easy to overlook these subtle human touches, all in the interest of getting the work done. We say that the work comes first and people's feelings will have to wait. We question why we have to take the time and energy to motivate them—"They're getting paid to do the work, aren't they?" is a common excuse. So we sacrifice attending to these daily needs of our people. But when that happens, little by little we fail to inspire, and it is no wonder people fail to care about their work or do their jobs conscientiously, following the standards, on a regular basis.

The Ultimate Human Potential: Creativity

All facets of Lean recognize the role of employee ideas and participation in the improvement process at every level of the organization, starting with the first-line operators who actually perform the work. But it is inherently difficult to change a culture where people are afraid to express opinions and ideas into one where those opinions and ideas are welcomed and rewarded. Countless companies have spent many years trying. While a company may profess that it wants people to freely suggest changes to the current system, what oftentimes happens is that supervisors, who wield the real power with employees, continue to resort to their positional power to force things to happen as they see fit. It may be that they simply don't know of any other way to get the work done than to force their own will on people and have it done "their way." Worse yet, even when supervisors do become involved and excited about making improvements, they become frustrated because they have not been taught how their management style must change to fit a culture of employee involvement.

But even when teamwork training and other leadership programs are enthusiastically applied to help supervisors overcome these weaknesses and to learn how to integrate operators into the daily management processes, we still find that ordinary workers, including those who have experienced participating in the kaizen-related activities, may still find it hard to contribute. It is not unusual to hear excuses like:

"I'm not an idea person."
"Mary is the creative one. Ask her."
"I haven't had a new idea for a very long time."
"I gave my idea last year. I'm done."

What holds people back from tapping into their innate ability for creative thinking, a blessing that all human beings possess? Dr. Robert Maurer, in his self-help book *One Small Step Can Change Your Life: The Kaizen Way*, suggests both the cause and the cure using fundamental concepts of kaizen as a prescription.

First, Maurer explains, any type of improvement implies change and change creates fear, especially when it is dictated to us from above and mandates substantial reform. People fear what the change will mean for them and the consequences it will bring to their current comfortable situation. This fear in turn engages a reaction in the primitive part of the human brain known as the fight-or-flight response, which is a survival mechanism that shuts down rational and creative thinking. This response makes sense when we consider, for example, being pounced upon by a saber-toothed tiger—we don't want to take a moment to ponder whether it would be smarter to run into the forest or to head for the hills; we simply run as fast as we can to avoid being eaten.

In our modern workday lives, we may not always need this survival mentality, but the way our brains are wired still creates this barrier to creative thinking. In their zeal to capture the benefits of Lean, companies oftentimes go out and, with good intentions, dictate to their people that they shall "go forth and improve," reaping huge rewards through the benefit of everyone's ideas. Unfortunately, this effort serves to create fear—fear of not being able to meet the high expectations of management or fear of adverse consequences—and the great promise of improvement never comes to fruition. If anything, people may simply give token suggestions or rehash ideas they have considered before as a way of placating management. But everyone winds up disappointed and resentful.

Maurer suggests that the kaizen philosophy of small steps toward continuous improvement is a way of sidestepping the fight-or-flight response and avoiding this conundrum. When people are allowed to take small steps toward improvement, they do not have to fear adverse consequences, and this keeps their minds open to creative ideas and thinking. These small but regular improvements, then, add up over time to big results. Furthermore, Maurer suggests that the way to get this process started is by asking small questions, which "create a mental environment that welcomes unabashed creativity and playfulness."*

> By asking small questions, gentle questions, we keep the fight-or-flight response in the "off" position. They allow the brain to focus on problem-solving and, eventually, action. Ask a question often enough, and you'll find your brain storing the questions, turning them over, and eventually generating some interesting and useful responses.
>
> Even if you're not an aspiring novelist, small questions can help calm the fears that squelch creativity in other realms of life. Consider how the microwave was invented, for example. Perry Spencer didn't sit around the house, drumming his fingers and pounding his forehead, thinking, "How, how, *how* can I invent a device that will revolutionize kitchens around the world?" Spencer, an engineer at Raytheon, was at work one day when he left a candy bar too close to some radar equipment. The snack melted, and he asked himself, "Why would radar have this effect on food?" This small question led to answers that led to other small questions whose answers eventually changed how you and I make dinner.†

The concept of asking questions with an open mind, like a child, was taken up by the original TWI Job Methods training manual from 1943, which stated, "The success of any improvement depends on our ability to develop a questioning attitude." The manual went on to explain how children learn about the world around them by asking questions and that, as adults, "we stop questioning

---

* Robert Maurer, *One Small Step Can Change Your Life: The Kaizen Way*, Workman Publishing, New York, 2004, 35.
† Ibid, 44, 47.

things too soon." The heart of the JM method, where the ideas are generated, is Step 2, Question Every Detail. Here we simply question each detail of the job, going through the six basic questions in a specific order:

Why is it necessary?
What is its purpose?
Where should it be done?
When should it be done?
Who is best qualified to do it?
How is the best way to do it?

The answers to these questions are, in fact, our ideas, and we write them down under the ideas column of the JM breakdown sheet as we continue to question all of the details of the job we are analyzing.

By asking these basic questions about minute details of a job, we can come up with ideas that, when put together into a comprehensive proposal, make up a step in the right direction toward continuous improvement. This step does not have to be a big one, but every step toward improvement is significant. No longer are we waiting futilely for some "big idea" to somehow "fix everything" and provide us with the breakthrough answer to all our problems. In fact, we no longer have to be afraid that we have no good ideas to give—we simply ask the questions and let ourselves find answers to them. In the absence of fear, our minds relax and we are, even without our realizing it, able to access that creative source of the brain that will now give us answers to these questions. These answers, it turns out, are in fact our unique and creative ideas.

When doing a JM exercise at a plant that refurbishes power generation turbine parts, a supervisor was questioning the process of having to crawl into a machine in order to remove a set of guards from the part just completed—only to have to place those same guards on the next part that would be processed. The machine blasted off the worn surface of the burner unit, and the guards were put on to protect certain delicate areas of the unit that would be damaged by the blasting. This was an arduous process and somewhat dangerous because of the confined space inside of the machine. The supervisor questioned, "Where should it be done?" and thought that the guards should be put on outside of the machine, where there was more space.

After some discussion with the class, the supervisor felt that they could change this procedure. Then they went on to the next question: "When should it be done?" The idea came up of putting on the guards for the next part while the previous one was still in process to save downtime on the machine, but they only had one set of guards. "Then why don't we make up another set of guards?" the supervisor asked incredulously. This simple Lean concept of making setup preparations while a machine is running was never applied to this process because the guards had always been removed and installed *inside of the machine*. Once a simple question was asked and the location changed, it became very easy to see the next improvement step, which wound up creating substantially more productivity on a bottleneck process.

Even after many years of energetic Lean activities, companies that take up this simple questioning process using the TWI JM method are astounded at how many more improvement ideas, simple but effective, they are able to generate. What is more, they are able to continue repeating the process because their people are no longer paralyzed by fear and cut off from their creative faculties. Having a method that recognizes the human element of where ideas come from, they are now able to fully realize the true potential of continuous improvement.

A Culture of Competence

In the example above about the Viet Nam veteran whose behavior went sour when he saw the hero treatment soldiers from the war in Iraq were receiving when they got home—a very different reception than veterans from his war got when they returned to the United States—we noted how in most cases like this untrained supervisors feel justified when they wind up firing employees who break zero-tolerance rules such as insubordination or harassment. When we point out that their behavior, though not correct, was nevertheless understandable considering the circumstances, they proclaim, defiantly, "But they deserved it!" That may certainly be the case, depending on the offense that led to the termination. However, this is still a failure of leadership. Because the supervisor did not get into the problem earlier, when it was smaller and before the person committed some "unforgivable sin," he or she allowed the situation to deteriorate and descend to its final outcome.

In Job Relations, we teach that leadership is about recognizing and dealing with problems while they are still small. By building strong relationships, supervisors can get to know each and every one of their people as individuals, and can spot changes in behavior or attitude before they become entrenched. When a problem is still small, and easier to handle, there will be a variety of possible actions that can be tailored to that person's unique needs and demands. In these early stages, the employee will not, as yet, have committed those critical infractions, which are done, after all, in order to get our attention because we *haven't noticed* the more subtle signals of dissatisfaction he or she has been giving so far. Moreover, with multiple options at our disposal, we will then have the opportunity to pick the right one that both solves the problem and meets our larger objectives.

There is a common misunderstanding that each person in an organization must be treated exactly the same for any given offense, as if people really were machines, and when something gets squeaky, a little oil always does the trick. Every person has his or her own motivation for the behaviors he or she exhibits, and what is a punishment for one may actually be a reward for another. Fairness is more about making sure each problem is resolved, not haphazardly or based on some simplistic formula, but by following a uniform method, as embodied by the JR four-step method. In other words, each problem is handled following a rigorous method of fact finding and evaluation of options. This method ensures that supervisors follow all practices and policies of the organization while, at the same time, allowing them to treat each problem individually based on the circumstances of each case.

In our JR classes, supervisors bring in real people problems they are facing and we practice using the JR method to find actions that help them get better results. We always begin by asking them what the objective for their problem is, and for the most part, it is some variation on "to make Johnny a better worker." When we get to the action they finally took, though, it is not unusual to hear, "We fired him."

"So, did your action meet the objective?" we ask.

"Not really," they say.

"What else could you have done," we challenge them, "besides firing him, that might have made him a better worker?"

In effect, this is about taking responsibility for one's people, which is the essence of competent leadership. As Stephen Covey likes to say, being responsible is about being "response + able," or able to respond to situations in a way that produces good outcomes. We like to call it, simply, competence. It is not enough to simply blame people for their errors or incapability. Supervisors must first ask what they are lacking that let the problem occur.

All three of the TWI skills are, in essence, about leadership. They develop specific skills that are targeted at leading people to more desirable outcomes. When employees have confidence in their supervisors to help them succeed and flourish in their jobs, it is a natural outcome that they will follow this guidance, knowing it will take them to that better place every individual strives for. If they follow you, then you are leading. Everyone wants their lives to improve day by day, and thus everyone is always looking for good leadership. But just having your heart in the right place is not enough. Competence in these skills is essential to gain the people's trust.

In this chapter we have seen how the TWI methodologies were designed to capture the human element in each of the skills they promote. By embodying a true respect for people, they get to the essence of leadership by teaching supervisors the skill of engaging their people when they perform the vital functions of teaching jobs, preventing and solving people problems, and improving methods to make them safer and easier to do. Once these skills are mastered, supervisors become competent in performing their vital role of getting out quality output at the proper cost. A focus on these skills, then, creates a culture of competence in which vital tasks, from both the supervisor's standpoint and the operator's standpoint, are performed skillfully on a regular basis.

Chapter 8

Culture Building at W. L. Gore

What kind of culture would a company have to have to be listed, for the thirteenth year in a row, on *Fortune Magazine*'s list of the "100 Best Companies to Work For" and to be thirteenth in that 2010 ranking?[*] Perhaps a company that *Fast Company* magazine called at the end of 2004, "pound for pound, the most innovative company in America,"[†] and ranked number thirty on the list of the world's fifty most innovative companies in 2009.[‡] W. L. Gore & Associates, the company that Wilbert (Bill) Gore started with his wife Genevieve (Vieve) in 1958 in their basement making electrical cable, has grown in its fifty-plus-year history to $2.5 billion in revenue with nine thousand associates. Today, they make thousands of diverse products, including fabrics (the waterproof, breathable GORE-TEX® fabric is perhaps their best known product), fiber, cable, medical devices, filtration products, sealants, and venting products. They are one

---

[*] 100 Best Companies to Work For 2010, *Fortune*, February 8, 2010.

[†] The Fabric of Creativity, *Fast Company*, December 1, 2004.

[‡] The Fast Company 50—2009, *Fast Company*, http://www.fastcompany.com/fast50_09/list-all.

of the two hundred largest privately held companies in the United States.[*]

Undoubtedly, a key factor in Gore's success has been the company's unique, nonhierarchical, team-based culture. In creating this type of work environment, Bill drew heavily from his experiences at the DuPont Company, where he worked for seventeen years. During his time there, he occasionally worked on small task forces, created to solve problems. He noticed that the small size of these task forces, and their lack of hierarchy, enabled them to problem solve quickly and effectively. The task forces also allowed Bill to better observe his colleagues' talents. These teams operated for short durations, until the problems that they attacked were resolved. Why, then, Bill wondered, couldn't he structure an entire organization to operate this way?

This thinking paved the way for Gore's flat "lattice" structure. Very few titles exist within the enterprise, and all employees are referred to as associates. The company is also free of traditional bosses and managers. Instead, Gore has leaders, who emerge based on their knowledge, skill, and ability to attract followers. Unlike bosses, they have no assigned authority. Every associate also has a sponsor—a fellow associate who upholds and teaches the Gore culture. Sponsors also help their "sponsees" maximize their contributions. Every associate has the freedom to change sponsors if they feel the relationship is not working.

Gore's pioneering corporate culture and management structure do more than simply create a favorable work

---

[*] W. L. Gore & Associates; History, Wikipedia, http://en.wikipedia.org/wiki/W._L._Gore_and_Associates.

environment. They play a critical role in driving business results, by fostering personal initiative, innovation, and direct communication.

At the heart of the culture are four guiding principles, which Bill formally outlined in a 1976 paper distributed to associates, called "The Lattice Organization—A Philosophy of Enterprise." These principles are intended to help guide associates in their day-to-day decision making, and they remain a key part of the company's culture.

Guiding Principles[*]

- **Freedom:** The company was designed to be an organization in which associates can achieve their own goals best by directing their efforts toward the success of the corporation; action is prized; ideas are encouraged; and making mistakes is viewed as part of the creative process. We define freedom as being empowered to encourage each other to grow in knowledge, skill, scope of responsibility, and range of activities. We believe that associates will exceed expectations when given the freedom to do so.
- **Fairness:** Everyone at Gore sincerely tries to be fair with each other, our suppliers, our customers, and anyone else with whom we do business.
- **Commitment:** We are not assigned tasks; rather, we each make our own commitments and keep them.
- **Waterline:** Everyone at Gore consults with other associates before taking actions that might be "below the waterline"—causing serious damage to the company.

---

[*] Taken from W. L. Gore website: http://www.gore.com/en_xx/careers/ whoweare/whatwebelieve/gore-culture.html.

The waterline principle uses a boat as an analogy to communicate the impact certain actions could have on the company. When someone drills a hole above a ship's waterline, it has relatively little impact. However, when someone drills a hole below the waterline, it threatens to sink the entire ship. This concept is put into practice when associates consider what is in the best interest of the entire company before making decisions on their own that might do damage below the waterline.

In effect, Bill was creating a culture where innovation could flourish without risking disaster. It was a combination of freedom and responsibility, so that individuals could thrive even as they committed their efforts to the long-term health and growth of the company.

Medical Products Division Focuses on Training Program Improvements

Gore's medical products division manufactures a large variety of medical devices, such as endovascular* stent grafts, synthetic vascular grafts, heart prostheses, and hernia repair material. With over 25 million devices shipped over their thirty-year history, the division prides itself in finding solutions to complex medical problems. Their products leverage Gore's vast expertise in fluoropolymer technology—the same technology responsible for products like GORE-TEX® fabric.

---

* The *American Heritage Medical Dictionary* (http://www.yourdictionary.com/medical/endovascular) definition: "Of or relating to a surgical procedure in which a catheter containing medications or miniature instruments is inserted percutaneously into a blood vessel for the treatment of vascular disease."

As with all of Gore's product divisions, quality is of utmost importance to the medical division. Doctors rely on Gore's medical technology to improve, and often save, patients' lives. Therefore, the products must perform in situations where failure is not an option.

To ensure that its medical products meet the highest standards, Gore follows very stringent quality control measures in the manufacture of these products. In 2008, the division launched a manufacturing training program designed to improve consistency among manufacturing trainers and associates, called the Manufacturing Training Solutions Project (MaTrS). The project mission was defined as "a team of associates dedicated to doing what matters—creating solutions that allow for a more consistent, effective, and transferable manufacturing training process."

Manufacturing trainers at each of Gore's medical plants teach jobs using approved standard operating procedures (SOPs). At the time the project launched, in order to improve training skills, some trainers also went through a train-the-trainer program designed to improve communication skills and elevate awareness of different learning styles. This program was helpful to trainers; however, it was recognized that more could be done to address how people learn jobs in different ways. In essence, Gore's medical division wanted more consistency with *how* process details were being taught to associates. Before a solution could be proposed, a comprehensive needs assessment was needed. Gore's learning and development team partnered with the manufacturing team to begin recruiting a person who could lead this training effort.

A Former Educator Takes the Lead

Suzanne Smith, who worked as an elementary school teacher for eight years before joining Gore, was conducting training and communications for divisional software implementations when she began transitioning into the new role of leading the Manufacturing Training Solutions Project. Suzanne's background in education enabled her to evaluate the validity of the different manufacturing training programs and methods she was reviewing.

In the meantime, a process engineer in Gore's medical division had attended an Association for Manufacturing Excellence (AME) conference where he had heard about the TWI program. He was excited about what he learned, and he shared these insights with manufacturing leaders, some of whom were also a part of Gore's operational excellence (OPEX) team, working on continuous improvement and Lean-related activities like kaizen events. After hearing and reading more about the TWI program, one plant began using parts of the TWI method at Gore, and they began seeing results. Suzanne informally heard about this effort from one of the manufacturing leaders who shared, "Here's something interesting that we're working on." Though the TWI activity was still at a very low-key level, she knew it was a method that had appeal to Gore associates.

After conducting considerable research, Suzanne discovered the Training Within Industry's Job Instruction (JI) program was the only well-defined and broadly used formal manufacturing training methodology. Of course, there were other options available, but they had been developed within certain companies, in response to the specific needs of those companies. With Gore's unique culture, Suzanne knew that she would need to find something that would fit in well with the way the company treated its associates—as team members and not as subordinates. Moreover, as a former teacher, she also knew that there was more to learning than just telling someone what to do. The program had to be based on sound principles of how adults learn and develop skills.

As she read literature on TWI, and began attending TWI webinars, Suzanne felt even more certain that the program would be a good fit for Gore. "It became evident that Gore's unique culture was a good match with the TWI philosophy," she says. In addition to being impressed with the program's time-proven success rate, which immediately gave it credibility, Suzanne saw many parallels between the TWI methods and Gore's culture. For example, the TWI methods focused strongly on leadership development, and this philosophy aligned with Gore's freedom principle, defined as "being empowered to encourage each other to grow in knowledge, skill, scope of responsibility, and range of activities." As President and CEO Terri Kelly said in January 2010 after the company made *Fortune Magazine*'s "100 Best Companies to Work For" list, "Gore remains true to its core values…. We recognize the importance of fostering a work environment where people feel motivated, engaged and passionate about the work they do."[*]

Suzanne decided to take on the role of leading the MaTrS project, and although she already had a keen

---

[*] W. L. Gore & Associates Continues Its Tradition as One of Nation's Best Workplaces, W. L. Gore Website: News and Events, http://www.gore.com/en_xx/news/FORTUNE-2010.html.

sense that TWI would be the tool she wanted to use, she pushed forward with the needs assessment and spent three months gathering data across the entire medical division. The results identified three improvement opportunities related to training:

1. Organizational training roles and expectations
2. Training consistency
3. New hire training

Suzanne felt she had good tools in place to meet all three of these identified needs head on. She then revisited JI as a potential training methodology to improve training consistency. She began passing out books and hosting webinars where people were invited to come and have a cup of coffee while learning about TWI. Since Gore's medical division had already tried TWI at one of its plants, it had some momentum built up for the program. But she didn't want to rush in and announce, "This is what we're doing." She felt it was much smarter to "be intentional, but be patient" and to build it up gradually.

What is more, Suzanne needed to establish a project core team and "get the right people in place right up front." First, she recruited a training coordinator that had already been using JI during Gore's initial introduction. He was a natural shoe-in to help out and was very supportive of the effort. Another training coordinator joined the team because she was very passionate about training and had been looking for something like JI. The final training coordinator joined the core team much later, after she had taken the initial ten-hour JI training and attended the TWI Summit in Cincinnati in May 2009. Her excitement and enthusiasm while attending these events made her a perfect fit to join the project.

In the meantime, Suzanne had contacted the TWI Institute in January 2009 and found they could give her the support needed to successfully learn how to conduct the TWI program at Gore. Based on the material she read about TWI, her conversations with the authors, and the medical division's prior work with the TWI program, she was able to construct an organizational training role straw man. She used this model to communicate her suggestions for manufacturing training roles and expectations to the rest of the medical division (see Figure 8.1).

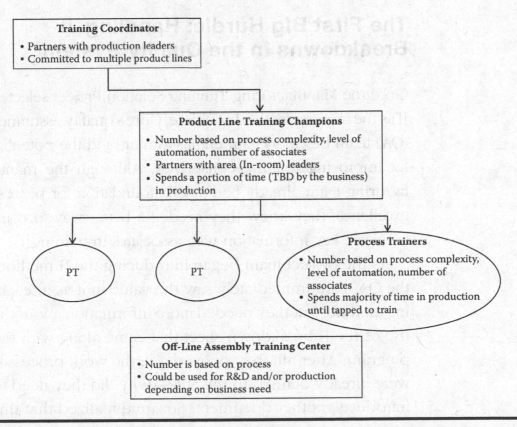

Figure 8.1 Manufacturing trainer organizational model—February 27, 2009. (Reprinted with permission of W.L. Gore & Associates, Newark, DE.)

She presented the straw man during a kaizen event where a cross-functional and cross-divisional team was gathered to tackle the task of refining the Gore training organizational model. Once the mission was accomplished, and a strong model was created by the kaizen team, Suzanne felt secure that the program was set up for success. This model would be the foundation for the successful implementation of JI throughout the initial pilot and rollout process, again ensuring that the right people—associates passionate about training—would be on board to support success.

The First Big Hurdle: Handling JI Breakdowns in the Quality System

Once the Manufacturing Training Solution Project selected JI as the training method of choice, Gore's quality assurance (QA) team began to gain enthusiasm around the potential for improving training consistency. Although the manufacturing team already had a solid foundation for process excellence, they knew they needed a better way to communicate key information with associates in training.

As the project team began introducing the JI method, the QA team immediately saw the value in it as a teaching method, but they needed more information about the use of the JI breakdown sheet that came along with the program. After all, they pointed out, the work processes were already defined in the SOP—why did they need to introduce another document? Suzanne realized that this concern was an important issue that would need to be thoughtfully considered.

She immediately set up monthly meetings with QA associates in order to get them involved in how the training method would be introduced in a pilot test (see next section). Her main goal was to get key QA associates and QA leaders involved in the JI pilot to ensure the implementation team was meeting all quality system requirements. These meetings opened up with questions like, "How do you think the pilot is going?" and "What have you seen in the pilot that works, and what have you seen that doesn't?" Conversation then moved quickly to the real issue: How could JI be integrated into the medical division's current quality system, and more specifically, how could the JI breakdown sheets be brought into the existing quality system to complement SOPs?

The SOP Is the Textbook, the JIB Is the Lesson Plan

When Patrick did the initial ten-hour sessions of JI for the group in Flagstaff at the end of May 2009, he explained that the JI breakdown sheets were like notes for trainers to use during instruction. Per ISO 9001:2008 standards, all documents need to be "subject to the document control requirements of clause 4.2.3."* And as QA pointed out, you can't have *uncontrolled notes* in the manufacturing areas." Creating new documents outside of the controlled SOPs would be counter to the foundation of the established quality system.

---

* International Organization for Standardization, http://www.iso.org/iso/iso_catalogue/management_standards/iso_9000_iso_14000/iso_9001_2008/guidance_on_the_documentation_requirements_of_iso_9001_2008.htm.

Suzanne began by explaining that they should look at the JI process at its most basic level—teaching an associate who has never done the job before. The SOP documents the process to be learned and routinely executed; training is the communication that creates a bridge between the technical process and quality execution of that process. And, there is always room for improving training, especially when your audience is a person who has no understanding of the work and how it is to be done. The trainer's job is to ensure an associate's work aligns with the details of the SOP. Going back once again to her experience as a teacher, Suzanne remembered how she would create lesson plans ahead of time that would guide her through her classes, making sure she covered the material in the right amount and the right order without missing anything. That, of course, would not take away the need for textbooks that her students would study on their own even as she guided them through the material in a live classroom setting.

Using that model, she explained how the SOP was analogous to the textbook: a detailed description, with pictures and other illustrations, of how a job is done that can be studied for overall understanding and to get information on minute details when needed. The JI breakdown sheet, on the other hand, is much like the teacher's lesson plan: an outline of what you are going to teach that is only for the use of the trainer in order to allow him or her to teach the job properly, in the correct order, and without missing anything. In the same way that students need both the textbook and the classroom instruction in order to master a subject, manufacturing associates need a trainer to help them master the job even as they study the SOP for overall comprehension. And because we are dealing with adults, the added dimension of getting the person to care about learning the job increases the need for strong instruction skill.

By changing the paradigm of the JI breakdown sheet from "notes" to "lesson plan," she was able to communicate how the concept of JI worked to effectively and consistently teach jobs to manufacturing associates. Moreover, by emphasizing the relationship of this lesson plan to the textbook, she was able to reassure them that, far from replacing the SOPs, the JI breakdown sheet would rely on the SOPs for their content and substance. Clarifying the fact that associates would still be required to use the SOPs for details and guidance during and after their training was a vital factor in obtaining QA endorsement in the use of the JI method by trainers within the manufacturing areas. (See Chapter 4 for a complete discussion on how JI breakdown sheets work with SOPs.)

What remained, then, was for the project team to figure out how to control these training documents on the plant floor.

Integrating JI Breakdown Sheets into the Quality System

In the meetings with the quality team over several months, Suzanne's team worked diligently to explore different means by which they could track and control the JI breakdown sheets in a way that would satisfy the quality system requirements. One concern was that older versions might be left behind, and the trainers would use these outdated versions, even after the corresponding SOPs had been updated or changed. And QA wanted assurance that trainers would consistently engage the right approvers when creating the JI breakdowns, and this might be more difficult if the documents remained outside of the established quality system. The project team considered different methods for approving, numbering, storing, and updating the JI breakdown sheets, but continually got hung up on the complexities of aligning these documents with the SOPs they were connected to.

As oftentimes happens when looking for a complex solution to a simple problem, they found that the answer was far easier to implement than expected. "We already have a system [for controlling SOPs]," a quality leader reminded the team, "[so] let's make the system work for us." Why come up with a different way of doing this? Take the existing system and make changes so as to integrate JI right into it. And, as it turns out, the changes needed to make this happen were minor.

A more formalized path toward successful integration of JI into the quality system began to emerge. Suzanne, and a QA leader committed to the successful integration of JI, next met with an expert on the division's product life cycle management (PLM) software team. From this meeting a solution was solidified that would make use of the already established bill of materials (BOM)/documents structure in PLM to establish a direct electronic parent-to-child relationship between the JI breakdown sheets and their related SOP.* Then, when you called up an SOP document in the software, you could view its BOM to see the associated JI breakdown sheets; because any one SOP covered the whole of a particular process, it

---

* Because of confidentiality agreements with W. L. Gore, we are unable to reproduce those screens in this book.

could consist of several breakdown sheets. Additionally, the software housed the quality change control process and was designed so that if any changes were made in the SOP, the applicable job breakdown was easily viewable to ensure potential updates would be considered by trainers. All original breakdowns put in the system were to be formally reviewed and approved per the quality system requirements.

Since the JI breakdown sheets would be governed by the established quality system along with the SOPs, there was no need to establish a new numbering system. The JI breakdowns would receive their own number within the established quality system. The JI breakdown sheet then had its own number and its name became JB #1 for xyz, with xyz being the number of the SOP. In that way, the title of the JI breakdown sheet referred directly to its related SOP. They included this information in their breakdown sheet format (see Figure 8.2), where the document number is the number assigned to the JI breakdown sheet and the JB number simply references how many breakdowns go with that SOP. For example, this document is JB number 3 of five total JIBs for this particular SOP number. Notice also the note at the top of each breakdown, which reminds the user that it is only in support of training and should not be used as a tool to actually build the product. This is in line with correct JI practice, which dictates that the breakdown is for the instructor's use and "not to be shown to the learner."

They also modified slightly the content of the job breakdown to include an SOP reference column that would allow the trainer to refer any of the important steps of the breakdown back to the correct step in the SOP (see Figure 8.2).

Document number: JB ######

Title: JB #1 for SOPXXXXXX

Note: This document is used to support a trainer's job instruction and is not be used to manufacture product.	
JB # X of X	Average time to train:
Significance of operation:	
Operation:	
Tools and materials:	

Figure 8.2   W. L. Gore job breakdown sheet format. (Reprinted with permission of W.L. Gore & Associates, Newark, DE.)

If trainers needed clarification or more detail on any one of the important steps being trained, they could quickly reference that information in the SOP. In this way, a link was made between the methods being taught by trainers and the standard work defined in the SOP. This JI breakdown template, along with an SOP defining the use of the Job Instruction training method, became documents within the quality system. With a means now to ensure that correct and up-to-date breakdown sheets that complied with the methods dictated by the SOPs were always used when

Important Steps	SOP Reference	Key Points	Reasons
A logical segment of the operation when something happens to advance the work	SOP Step Reference	1. Make or break the job 2. Injure the worker 3. Make the work easier to do	Reasons for the key points
1.			
2.			
3.			
4.			
5.			
6.			
7.			
8.			

Figure 8.2 (continued).

training jobs, the path forward to use the JI method was communicated to QA and manufacturing leadership.

Suzanne felt that the ease by which they could update the PLM software within the established quality system to accommodate the JI breakdowns "spoke to the design of the quality system's tools." It also shows how the JI breakdown sheet fits seamlessly into a sound system of quality control. In the end, though, what was most important to the success of the MaTrS project was that she was able to communicate this critical piece and demonstrate it—they had made a mockup of the system to show how it would work—thus gaining the support of the key stakeholders to move forward with using JI breakdowns. Being able to communicate this functionality and control of the JI breakdowns within the quality system was the key to implementing improved training consistency.

Rollout of the Pilot

When Suzanne determined that TWI was the right tool to support the second MaTrS project deliverable, she began planning with the TWI Institute to run two ten-hour sessions of JI in late May 2009. She recruited twenty learners and some fifteen observers. As qualifications to participate as a learner in the class, associates had to:

- Be from a plant with organizational training roles and expectations
- Be currently fulfilling a commitment as training champion (title for job trainers)
- Have excellent written, verbal, and interpersonal communications skills
- Have the ability to organize, schedule, and facilitate meetings
- Have the ability to identify the following about processes in their area:
  - Most complex
  - Longest training time
  - Priority of production and training tasks

After the training was completed, these JI-certified trainers then went back to their respective plants, where they selected jobs to break down and train using JI in order to measure the effects of the training method. Suzanne's team worked with each plant to establish unique criteria for selecting initial jobs to pilot the method. For example, at the plant where they had already been using TWI, they selected two jobs, each in two different areas, in order to compare them to what they had done previously with JI. In two other plants, a total of eight people had taken the JI class, which allowed some twenty-five jobs to be broken down based on the following criteria:

- Complexity
- Training time required
- Number of new associates

Each plant proceeded with the pilot at a different pace but continued to move ahead because of teamwork and the support and coaching they gave to each other as budding JI trainers. The members of the MaTrS team, who led the activities at each plant, as well as the training champions leading the pilot training, helped each other out, both physically and emotionally, even going to other members' meetings and helping them with training demonstrations. As Suzanne explained at the end of the project, "Because our core team had the support of each other, we continued to stay on track. One person could help another when help was needed." Each associate involved knew he or she was stepping into uncharted territory by using training techniques new to the company. They had to get it right. But because of this mutual team support, as well as strong support from leadership, each individual could have faith that there was someone there to help and make sure he or she did not fail.

Introducing the Program While Learning to Make Job Breakdowns

In the beginning, and this is true for any company starting out using the JI method, trainers work hard making job breakdowns. While the use of the four-step method for training is pretty well understood and mastered in the initial ten-hour training sessions, much more practice and experience is needed to perfect the skill of making good job breakdowns. By working together and trying new things, members of the MaTrS team came up with a four-part development procedure, much like the JI four-step method itself, in which they took their new job breakdowns through a rigorous process of making sure they had the important steps and key points exactly as they needed them. One training coordinator in particular became a key driver of the development of this process within her plants. This process delivered multiple benefits to the trainers and to the company overall because it not only helped them arrive at good job breakdowns, but it brought various functional groups together to discuss creation of the breakdowns even as the trainers gained essential practice in using the JI four-step method to teach the jobs.

In the first round of practice, remembering what Patrick had told them in class about trying out and "road testing" a new breakdown on the floor, they took a rough draft of the breakdown and taught it to someone in the plant. Because they were conscious of not slowing down production in the initial trial runs, they decided to have area or production leaders act as trainees and learn the job. These leaders did not have to be from the area or even know about this job. It was a way of seeing if the breakdown made sense to a knowledgeable associate, and it allowed for a fresh set of eyes to look at the breakdown to see if it worked. Over and above that, though, the method also created the added benefit of introducing the JI method to these leaders as they participated in the learning process. This effort went a long way in creating interest and momentum in the JI program as, gradually, more and more manufacturing team members were exposed to the program and became enthusiastic about its use. Suzanne saw this activity as a ground-up process that fit in well with the Gore culture.

With some feedback from this initial trial, the trainers made adjustments to their breakdowns and then set out to road test them again, in the second round, with process engineers acting as the learners. Suzanne realized the benefit of engineer involvement with JI right up front because she knew that engineers are key process and SOP stakeholders. Engineers, who often write the SOPs, typically use technical terms to define the process. As a result, new learners that are training to the process may need further explanation. This second round allowed manufacturing trainers and process engineers to partner closely to discuss JI breakdown content, and allowed trainers to verify that their breakdowns were teaching the job correctly as directed by the SOP.

By this point, the trainers had gotten their breakdowns technically correct, and they were ready to introduce them to the *gemba* and to those who actually performed the work to be trained. In the third round they trained the jobs to the leaders of the area who were directly in charge of these jobs. This allowed them to get feedback on how the breakdowns were written, on things that were left out or missing, or on things that were difficult to understand. This also allowed the area leaders to understand in more detail the training that would be conducted with their teams.

In all three of these initial rounds, while the main objective was to fine-tune and perfect the breakdown for the job, each time the trainers practiced using the Job Instruction four-step method as they taught the job to leaders and engineers. Members of the MaTrS team served as coaches and assessors, giving trainers feedback on their performance and helping them perfect their use of the method. Moreover, people serving as learners or observers could witness the JI process in action and get a good feel for what this new training program was all about.

In the final, fourth round of the breakdown development process, a manufacturing associate who did not know the job would be trained. Here the assessor would observe the instruction but say nothing and do a final evaluation of the trainer's use of the JI four-step method. The trainers worked hard at sticking exactly to the four steps, knowing they were being assessed. In effect, this was the final test to be sure the breakdown was understood and working as designed. But it was also a final check of the trainer to be sure he or she had mastered the JI four-step method before he or she went out onto the floor to begin the general training of the job.

Once the job breakdowns were completed and ready for use, the trainers took them out to the floor and began training people, both new and experienced associates, who actually performed these jobs. These training sessions, in effect, constituted the "pilot" of the JI method, and the team was sure to take note of any metrics associated with each of these jobs for use in the final evaluation. As the actual training moved forward, one training coordinator noted how the JI four-step method automatically took into account all three of the training styles they had been taught in their previous train-the-trainer program: auditory, visual, and kinesthetic. Whereas previously trainers would try to adjust to whichever style the learner exhibited, with JI they were covering *all three styles*, regardless of who the trainers were, when they "told, showed, and illustrated" the details of the job in Step 2, and then actually had the learner perform the job, the kinesthetic* approach, in Step 3.

In summary, the four-stage process of making a JI breakdown went as follows:

First round: Taught to training champions not familiar with the process to check general understanding of breakdown content; revisions made to breakdown.

Second round: Taught to engineers to confirm the technical content of breakdown in relation to the SOP; revisions made to breakdown.

---

* Webster's Online Dictionary (http://www.merriam-webster.com) definition: "Of or related to kinesthesis: the ability to feel movements of the limbs and body."

Third round: Taught to production leader in charge of the job being trained in order to check for missing content or difficulty in understanding; revisions made to breakdown.

Fourth round: Taught to manufacturing associate unfamiliar with job as final check of breakdown and formal assessment of JI trainers.

Getting Engineers Involved in the Process

One very important benefit the project team noticed while conducting the pilot was increased communications between engineers, leaders, and trainers. One training coordinator observed, "It was great to see these groups work so well together." While the engineers knew all about the equipment, specifications, and technical details of the job, working on the JI breakdowns allowed more best practices to be captured to enhance training to the job. The manufacturing steps might entail a great deal of dexterity in how the materials and tools are handled and these could then be identified as key points in Job Instruction.

Serving as learners in the practice instructions, the engineers brought up questions about the jobs and, by looking at the jobs from the learner's perspective, they quickly began to see things that could be changed or improved. This added to both the quality of the breakdowns and the results achieved after the training was completed.

Engineers could then refer back to their Job Instruction experience as they worked on other SOPs and new projects. This was borne out in one plant where a production leader who had taken the JI class returned to work on a new product development effort. Working with engineers to design new processes, including writing SOPs, she incorporated the JI technique into the work they were doing, and it began to show positive impact almost immediately. By enhancing the development process to include key points of the work, they were able to bring the engineering work that much closer to the *gemba*.

Results and Conclusion

Throughout this pilot period, Suzanne and the project team had been meeting regularly with QA, as described in the previous section, to work through how the JI method, and in particular the JI breakdown sheets, could be integrated into the existing quality system. Suzanne knew that these discussions were less about how the pilot was being conducted and more about being ready to launch and sustain JI after the pilot was over. The original goal of the project was to deliver a training method to the division that could meet the project goals, and she had confidence that the pilot would prove JI to be the effective tool they needed. Her work with the key stakeholders, then, even as the pilot was being conducted, began to pave a path for that method to be implemented seamlessly once the pilot was complete.

As the end of 2009 approached, her team collected the results of all the metrics and data from all the jobs being taught using JI, as well as feedback from the key people involved. These data made up the pilot summary report, which was added to the original JI pilot quality plan and included the following:

- Any deviations from the original plan
- A list of associates who participated on each plant's implementation team
- Associates who had been identified to potentially become JI master trainers
- Templates that had been developed during the pilot to aid in JI, such as the training timetable, job breakdown template (see Figure 8.2), job breakdown task list (used to prioritize and track JB creation), and trainer/JB assessment form (see Figure 8.3)
- A summary of all JI activities in each plant, including the SOPs that were broken down, the rationale for introducing JI in that area, and any associated activities
- Qualitative feedback collected by survey from learners that went through a training session using JI
- Recommendations on incorporating JI into the quality system
- Recommendations on how to assess JI trainers and JBs
- A conclusion statement that indicated the requirements of the quality plan had been met and the determination of the pilot as a success

This revision to the quality plan was submitted to key functional leaders for review and approval. Once the plan was approved, the results were shared with the MaTrS sponsorship team (same as a steering committee comprised of business leaders). With the addition of the summary report, the quality plan in essence became an implementation proposal outlining the recommendations on how JI could best be implemented throughout the organization. This allowed business leaders to then

General	Operation:			Start time:	End time:
	MD/JB number:	Date:		Plant:	Area:
		Learner:			
	Trainer:	O Engineer	O New associate	O Leader	Assessor:
	Comments:				

Setup	O SOP	O Blue card	O Job breakdown	O Workbench
	Comments:			

Prepare the worker	O Puts person at ease	O States job	O Finds what is known	O Interest in job
	O Reads SOP	O Reviews tools	O Reviews parts	O Puts learner in position
	Comments:			

Present the operation	O Important steps OOOOO	O Key points OOOOO	O Reasons why OOOOO
	O Clear instruction	O Minimal words	O References SOP
	Comments:		

Try-out performance	O Important steps OOOOO	O Key points OOOOO	O Reasons why OOOOO
	O Clear instruction	O Minimal words	O References SOP
	Comments:		

Follow-up	O Important steps OOOOO	O Key points OOOOO	O Reasons why OOOOO
	O Clear instruction	O Minimal words	O References SOP
	Comments:		

Job breakdown	O Minimal wording	O Minimal important steps	O Minimal key points per step
	O Follows SOP	O Reasons are clear	O Balanced
	Comments:		

Figure 8.3   Trainer/Job Instruction Assessment Form. (Reprinted with permission of W.L. Gore & Associates, Newark, DE.)

make a knowledgeable decision on whether JI should be rolled out throughout the rest of the division.

After careful review, the sponsorship team agreed with the recommendations and supported the divisional rollout of JI. What happened next is that the successful results of the pilot were summarized and communicated to the rest of the business so the business leaders could make their own decisions on when and how to begin implementation. Each business or plant is now making plans for leading its own JI implementation, and the level of enthusiasm is high. The role of Suzanne's team was to create a template for how to roll out JI and, having done that, she says, "The road is now paved, and each team is driving down it."

Learnings and Recommendations

Going through a JI pilot allowed the team to create an opportunity for a variety of learning experiences that they were then able to put into the recommendations for JI implementation on a broader scale. With the support and power that the small core team provided, in addition to the safety net cast by engaged business leadership, many opportunities arose to jump in and try out different ways to integrate JI. This ultimately led to a successful cultural, business, and quality system integration. Suzanne felt that what they learned along the way would certainly also benefit anyone trying to implement JI in a similar situation where you have a large organization that has a strongly defined culture. Some of her insights were as follows:

■ Begin with a strong trainer organizational model that you can use as a foundation for a successful implementation. Consider this as your foundation, upon which you will build your house, or as Jim Collins states, "first who, then what." First get the right people on the bus, and then decide where you are going to go.

■ Start slow and don't rush—baby steps. Build momentum by consistently sharing relevant information and effective communications with key stakeholders. Be patient and let people get involved naturally in the process without forcing them. Provide hands-on experiential opportunities with TWI.

■ Have a strong champion of the TWI program who can engage the appropriate stakeholders within your organization (like QA, manufacturing, and engineering). Depending on your organization, it is most likely better to delay your implementation and wait for a passionate champion to emerge than to start early without a strong leader and potentially have your efforts fail. This person must be well respected by the business and have good facilitation skills. In other words, he or she has to appreciate the complexities of what is trying to be accomplished in managing a project of this scale, know the needs of the business, and be passionate about training.

■ Get assistance or expertise in communications. Don't underestimate the behavior-changing power of good communications and marketing. Communications early in the process to show a vision for how JI can integrate into your corporate environment is a critical piece in getting the program off the ground and moving forward. Suzanne feels, "You simply have to spend the time up front on this to get people on board."

The issues faced by Suzanne and her project team to integrate TWI into an already well-established corporate culture proved challenging but not insurmountable. Their dedication and tenacity to pursue a cause they believed in with a tool they could count on were the most important factors to their success. Having accomplished the project's goal to deliver a training method, Suzanne's final task was to bring in the TWI Institute to train six of their key JI trainers to teach the JI ten-hour class. Chosen from the trainers that participated in the pilot, this group of trainers could then internally support the continued rollout of JI in the business at a pace business leaders could feel comfortable with. The JI master trainers graduated in early January 2010 and are now conducting JI classes following the implementation plans of each plant as they begin to make TWI a part of the way they do their business.

Chapter 9

TWI Returns to Healthcare at Virginia Mason Medical Center*

Virginia Mason Medical Center (VMMC) is one of the three major hospitals of Seattle with a staff of four hundred employed physicians and a total of five thousand employees. The hospital, which includes a main campus in the Pill Hill neighborhood overlooking Puget Sound and seven regional centers, has 336 beds and generated $650 million in net revenues in 2008. Most significantly, VMMC has long been recognized as the premier user of the Toyota Production System (TPS) in the healthcare industry, advising such venerable providers as the Mayo Clinic and the British National Health Insurance system on the application of Lean in healthcare. In 2008, they formed the Virginia Mason Institute (VMI) to provide education and training in the Virginia Mason Production System (VMPS), as they call their version of TPS, to other healthcare providers and organizations.

---

* *Note:* Improvement metrics depicted in this case study are a snapshot in time and results today may differ.

VMMC took its first steps to becoming a Lean hospital in June 2002 when it sent its entire top level of management, over thirty people in total, on a thirteen-day study tour of Lean manufacturers including Toyota in Japan. During the study, this group of executives actually experienced hands-on work in the assembly lines in order to understand the concepts and application of standard work. Following these initial steps, the hospital began implementing TPS principles and tools, including Takt time, 5S, *heijunka*, and the *kanban* system, and in the first four years of application saved $6 million in planned capital investment, freed up 13,000 square feet of space, reduced inventory costs by $360,000, cut walking distances, shortened bill collection time, slashed infection rates, and, most importantly, improved patient satisfaction.[*] By the end of 2009, those results increased to $11 million saved in planned capital investment, 25,000 square feet of space freed up, $1 million saved in inventory, and staff walking distances reduced by 60 miles per day.[†] What is more, they were able to cut down by 85% the time it takes to get lab results reported back to patients while reducing labor expenses (overtime and temporary labor) by $500,000 in just one year.[‡]

In 2007, as part of their ongoing and energetic effort to replicate the Toyota system, they read *Toyota Talent* and realized for the first time that there was a method they could learn that would help them promote their efforts at standard work, a foundation of TPS. They

[*] Toyota Assembly Line Inspires Improvements at Hospital, *The Washington Post*, June 3, 2005, p. A1.
[†] Virginia Mason Medical Center, 2009 VMPS Facts, www.virginiamason.org/home/workfiles/pdfdocs/press/vmps_fastfacts.pdf.
[‡] Ibid.

immediately began using the concepts they learned from *Toyota Talent*, making breakdowns of jobs such as "Hand Hygiene in Compliance with CDC and WHO Hand Hygiene Guidelines" and "Time Out Prior to Surgery or Invasive Procedures," and trying to teach these jobs using the TWI method. They found out, though, that their breakdowns (see Figure 9.1) were too detailed and learners were confused by the training process—just the opposite of what they expected to find from this time-tested method so effectively used by Toyota.

The head of the Kaizen Promotion Office at VMMC, Linda Hebish, attended a TWI workshop given by us, the authors, at an Association for Manufacturing Excellence conference in San Diego in June 2008. There, she learned that the TWI programs were traditionally trained following a well-defined recipe that ensured proper learning and use of the methods. Moreover, she found out these training courses were packaged and ready to teach and could be turned on literally at a moment's notice. She became convinced that TWI could help advance the ongoing kaizen efforts at VMMC, but needed an enthusiastic promoter who could lead this part of the effort. When she got back to Seattle, she worked with the executive leadership team and then requested for reassignment consideration one of their nurses from oncology, Martha Purrier, to lead the effort to bring these training programs to VMMC.

The Right Person for the Job

When she got the call, Martha Purrier was the director of the inpatient unit of VMMC's oncology department. As

Job Breakdown Sheet

Date: January 21, 2008	Team leader: Donna S., Gillain A., Joan C.	Sponsor: Donna S., MD
Area: All areas providing direct pt care or in contact with pt care supplies, equipment or food	Job: Hand hygiene in compliance with CDC and WHO hand hygiene guidelines	Written by: Joan C.
Major Steps	Key Points	Reasons for Key Points
Step 1: Identify the need for clean hands	Remove artificial fingernails or extenders when in direct contact with pts or their environment	Artificial nails house germs that can be passed on when you touch pts
	Clean hands whether or not you use gloves (i.e., before putting on gloves and after removing gloves)	Gloves are not a substitute for cleaning hands because gloves don't completely prevent germ transmission
	Before direct contact with pt, pt's environment, or equipment	Protect the pt against harmful germs carried on your hands
	After direct contact with pt, pt's environment, or equipment	Protect yourself and the healthcare environment from harmful pt germs
Step 2: Inspect your hands to determine best cleaning method	If not visibly soiled, use alcohol-based gel	Cleaning with gel is faster, more effective, and better tolerated by your hands
	Visibly soiled hands or hands with fecal contamination require washing with soap and water	Dirt, blood, feces, or other body fluids are best removed with soap and water (C. diff spores are not killed with alcohol-based gel)

Step 3: Use enough product to cover all hand surfaces and fingers	Gel: Cover all surfaces with a thumb-nail-sized amount	Friction and skin contact are required to remove germs
	Wash: Wet hands with water, wash with enough soap to cover all hand/finger surfaces	
Step 4: Spend enough time cleaning your hands	Gel: Vigorously rub until product dries on your hands	Antiseptic is not complete until fully dried (approx 15 seconds)
	Wash: A minimum of 15 seconds (the length of singing "Happy Birthday to You")	As least 15 seconds is needed to ensure complete coverage of hand surfaces
	Use paper towel to turn off water faucet	Prevent transfer of germs from faucet onto clean hands
Step 5: Let your hands completely dry	Moisturize hands with lotion available through central supply	To minimize contact dermatitis without interfering with antimicrobial action
	Put on gloves after hands are dry	Skin irritation may occur if moist hands come in contact with glove material
Step 6: Perform task with clean hands	Task is done immediately after cleaning hands	You may be distracted and touch unclean surface with clean hands

Figure 9.1 Original job breakdown sheet for hand hygiene. (Reprinted with permission of Virginia Mason Medical Center, Seattle, WA.)

a certified oncology nurse, she was also a teacher and instructor at various nursing schools in Seattle. While she was not looking to move out of oncology or her role as a nurse, she was excited about the work being done to date with TPS and immediately saw the application of good training when told about the TWI methods.

Martha was well known at VMMC as energetic, motivating, and tenacious around procedures. In her work with the Virginia Mason Production System, Martha had admired greatly the role of standardized procedures as a foundation for the improvement the organization so greatly desired. She found herself regularly quoting Taiichi Ohno, founder of TPS, when he said, "Without standards, there can be no improvement." Then she found herself repeating her own mantra around standards and kaizen: "Standards provide stability. Stability provides visibility. Visibility provides targets for kaizen." This was her favorite saying and the words for which she wanted to be remembered by.

With this strong conviction around standard work, Martha accepted the challenge of becoming a director in the Kaizen Promotion Office (KPO) with the mission of implementing the TWI program into the organization. She immediately began consulting with the TWI Institute on good strategies for getting the true value out of a TWI introduction and began benchmarking with other successful company rollouts. She organized the first TWI ten-hour sessions to be held in March 2009 and directed by TWI Institute master trainer Richard Abercrombie. They decided to select an initial group of ten who would take both the Job Instruction and Job Relations modules in the same week. These ten people included Martha and two others from the KPO, two staff from clinical education, two nurses, and three patient care technicians (nurse's assistants).

VMMC's first strategic move, then, was to begin implementation at the patient care technician (PCT) level with assistance from the clinical education office, whose mandate was to be sure all hospital personnel were well trained in their jobs. This move would prove extremely effective because, once the actual training of jobs began, nurses saw the positive effects of how the PCTs were being trained and came to Martha asking how they could receive the same training. Notably, the head of nursing, Charleen Tachibana, saw the power of the training method and quickly became a vocal advocate for its use throughout.

Initial Training and Insights

In preparation for their first TWI training, Martha instructed the ten participants to read *Toyota Talent* and review the TWI Institute website, which explained the different facets and history of the TWI program. She also had them read articles by Art Smalley on basic stability and Jim Huntzinger on why standard work is not standard. They had also been given an introductory letter on how to prepare for the training, which included bringing in a real job to practice on in the class. So Martha required them to observe key jobs in their areas and review the relevant standards.

As they approached the training, Martha and her colleagues were still not sure which hospital jobs would be appropriate for the training or for use with the JI method. Just about this time, as she was doing a purposeful literature search on the subject, going back deep into the nursing archives, she received a copy of an article from *The American Journal of Nursing* dated June 1946. It was written by Olive White, RN, who had worked at King County Harborview Hospital, just a few blocks down the street from VMMC in Seattle. In the article, which explained how TWI was brought to hospitals during the critical years of the war when the training of auxiliary workers was vital to compensate for the decrease in nurses, was a long list of trainable duties divided into four categories: (1) housekeeping duties, (2) transportation and communication, (3) patient care, and (4) clerical duties. The jobs on the list, tasks such as clean sterilizers, get supplies, assist with patient transfers, dress and undress patients, pass bedpans, test urines for sugar, give bed baths, chart stools, and fill in admissions and discharges, were not all that different from their current tasks in spite of the passage of over six decades. Martha felt invigorated to get this help from the past and to be able to "stand on the shoulders" of the nurses who came before her.

For the ten-hour JI class, then, they were instructed to bring in small jobs that could be done in the training room. The ten members decided on the following tasks:

■ Hand hygiene
■ Hand washing
■ Six-point hourly rounding
■ Collecting a specimen

- Blood glucose monitoring
- Removing a saline lock
- Donning and removal of gown, gloves, mask (PPE)
- Placement of patient ID band
- Stool occult blood testing
- Emptying an ostomy bag

During the ten hours of training, they found out that not only did these jobs seamlessly fit into the JI format for training, but they could indeed be trained more effectively than the way they were presently being taught.

Once the class was complete, Martha took the list of ten jobs they had worked on and presented it to the managers of the two inpatient departments that participated in the training. These managers felt that ten jobs were too many to start out with and suggested the list could be slimmed down to just two or three jobs that were the most critical or would have the most impact. Then these jobs could be taught as an initial pilot of the TWI program to demonstrate the value of the method. They helped reduce the number down to three jobs: (1) hand hygiene soap/water, (2) hand hygiene gel, and (3) hourly rounding.

Even after the ten hours of training, it was felt that the group still needed practice and that the breakdowns they made in the class could be further improved. So Martha divided the members into groups of three and had each group work on one of the jobs, herself being the fourth member of each group. The groups practiced doing the jobs and refined their breakdowns. Once this work was complete, they set up a room with stations for each job and hung the breakdown for each job on a wall nearby for easy viewing. A member of each group then taught the job to a member of the other two groups, and they went around the room from job to job until everyone was trained in the jobs of their fellow groups. When issues or insights came up during these practice rounds, they were written down on the breakdowns on the wall, further refining their breakdown sheets. They were able to evaluate how smoothly the training went, and when it was all over, all members of the team were trained in and able to do all three of the jobs.

Now they were ready to begin showing the hospital the power of this new training method.

Hand Hygiene: The Right Place to Start

From the very beginning of this effort, when they began reading *Toyota Talent* and found the TWI training program, the VMMC staff knew that hand hygiene would be an area they wanted to pursue. In fact, throughout the world it is common knowledge that washing your hands, and washing them well, is one of the most effective ways of staying healthy. This is true in the public at large, but much more important for the working people of hospitals and healthcare facilities, where infections can be transmitted from one patient to another. At VMMC, the slogan is: "The single most important thing we can do to keep patients and ourselves safe is hand washing."

Typically, the way to address this issue has been to create a hand hygiene campaign that promotes frequent and consistent hand washing using speeches,

PowerPoint presentations, posters, clear directions, and other awareness raising techniques. At VMMC they followed directives from the World Health Organization (WHO) that offered detailed instructions on how to wash hands, including simple diagrams of hands being washed (see Figure 9.2), and directives on when they were to be washed:

1. Before patient contact
2. Before aseptic (infection prevention) task
3. After body fluid exposure risk
4. After patient contact
5. After contact with patient surroundings

They even adopted the WHO pledge of "We will clean our hands before and after each patient contact and remind others to do so as well."

In spite of these valiant and sincere efforts at VMMC and healthcare facilities everywhere, audits of hospitals nationwide show that healthcare workers and professionals are not adhering to these important guidelines. A 2000 study in the medical journal *Lancet* showed that only 48% of general hospital staff were washing their hands as directed.[*] More recent data from the WHO claims that worldwide, on average, "healthcare workers fail to clean their hands 60% of the times they should when dealing with patients."[†] Audit data from VMMC itself showed 83.5% of staff washing their hands when

---

[*] D. Pittet, *Lancet*, 356, 1307–12, 2000.

[†] World Health Organization, *SAVE LIVES: Clean Your Hands, A Briefing Kit to Advocate for Action on 5 May 2010*, www.who.int/gpsc/5may/resources/slcyh_briefing-kit_website.pdf.

Handwashing Technique with Soap and Water

0	1	2
Wet hands with water	Apply enough soap to cover all surfaces	Rub hands palm to palm

3	4	5
Right palm over left dorsum with interlaced fingers and vice versa	Palm to plam with fingers interlaced	Backs of fingers to opposing palms with fingers interlocked

6	7	8
Rotational rubbing of left thumb clasped in right palm and vice versa	Rotational rubbing, backwards and forwards with clasped fingers of right hand in left palm and vice versa	Rinse hands with water

9	10	Duration of the entire procedure: 40–60 sec
Dry thoroughly with a single use towel	Use towel to turn off faucet/tap	...and your hands are safe.

Who Guidelines on Hand Hygiene in Health Care (advanced draft)/Modified According to EN 1500

Figure 9.2 World Health Organization hand hygiene instructions. (Reprinted with permission of Virginia Mason Medical Center, Seattle, WA.)

needed over the last year, up to the third quarter of 2009. The WHO also reports that 5 to 10% of patients in modern-day healthcare facilities in the developed world will come down with one or more hospital-related infections, the most likely culprit of which is healthcare workers not washing their hands and carrying germs from one patient to another.[*]

With these life-or-death implications, Martha set out to have her group tackle this important training issue: How can we get hospital personnel to not only wash their hands properly, but to do so on a consistent basis? What is more, with the onset of the H1N1 (a.k.a. swine flu) virus in the spring of 2009 and the uncertainty of an effective vaccine being developed in time for the coming flu season, the imperative for them to get this training done effectively could not have been higher.

In considering how to get the job of hand washing standardized throughout the organization, Martha remembered the words of VMMC's *sensei* Chihiro Nakao on the content of standard work. Mr. Nakao, founder of Shingijutsu USA, who were conducting the kaizen training at VMMC, had said the instructions should be "down to the hand motions." She considered how simply telling people that washing their hands was important or putting up posters showing people how to do it did not correct the problem to the levels needed. Here would be the true test of good job instruction technique, to both show and tell people how to do the job on a one-on-one basis in a way that would explain why they had to do the job

---

[*] World Health Organization, Health Care-Associated Infection and Hand Hygiene Improvement—Slides for the Hand Hygiene Co-ordinator, PowerPoint presentation, slide 6.

as instructed. By getting down to the hand motions, they would try to instill a deep understanding of each step of the process.

Having failed earlier in trying to use JI before they took the formal TWI Institute training, they now realized how their original breakdowns were too complicated and cumbersome for effective training (see Figure 9.1). They had to make them clearer and more concise. So they began by splitting the hand hygiene task into two separate jobs: one for washing with soap and water and the other for cleansing with gel. The two tasks were similar but contained distinct differences—gel is to be used when the hands are not visibly soiled with dirt, blood, feces, or other body fluids, while soap and water is needed to clean these contaminants.

Then, upon deeper reflection, they realized that their original breakdown was more a generalized set of work instructions that tried to describe everything that happened in the process. In the JI training class, they learned that they should stick to short and simple terms as they demonstrated the job, focusing on just those key points that were not readily seen in the demonstration (see Figure 9.3). More significantly, they found that by looking intensively at "the hand motions" of the job, they were actually able to explain the process in more detail even as they reduced the breakdown to just a handful of words.

For example, their original breakdown described the process of washing as "Wet hands with water, wash with enough soap to cover all hand/finger surfaces" and then to continue "a minimum of 15 seconds (the length of singing 'Happy Birthday to You')." After doing the job and studying the process referring to the WHO

Version 1.0
07.07.09

Job Breakdown Sheet

Operation: Hand hygiene washing

Parts: Soap, running water, disposable towel

Tools and materials:

Important Steps	Key Points	Reasons
A logical segment of the operation when something happens to advance the work	Anything in a step that might: 1. Make or break the job 2. Injure the worker 3. Make the work easier to do, i.e., knack, trick, special timing, bit of special information	Reasons for the key points
1. Wet hands	1. Without soap	If soap first, it rinses away
2. Apply soap	1. Cover surfaces	
3. Rub hands	1. Palm to palm 2. Palm to backs	
4. Run fingers	1. Thumbs 2. Interlocking 3. Backs of fingers to palm 4. Tips to palm	
5. Rinse	1. Leave water on	
6. Dry	1. Use towel to turn water off	Prevent recontamination of hands

Figure 9.3   New job breakdown sheet for hand hygiene. (Reprinted with permission of Virginia Mason Medical Center, Seattle, WA.)

diagrams, they got into more specific detail with the procedure. First, they made it clear what should be put on the hands first, the soap or the water. This was a source of considerable variation as many people found it more efficient to put soap on their hands first and to begin washing with water directly. However, soap lathers more completely when put on top of water and will tend to rinse away when applied the other way around. So the "efficient" way of washing hands was actually diminishing the effect and quality of the work. The hands should be thoroughly wet before applying soap, and this was clearly taught in the new breakdown:

Step 1. Wet hands.
Step 2. Apply soap.

Next, their original breakdown simply specified that "all hand surfaces and fingers" needed to be covered with soap because "friction and skin contact are required to remove germs"—the reason for the key point. This was clearly not "down to the hand motions" and left too much ambiguity over how to make sure all germs on the hands were cleaned off. The next two steps of the new breakdown specifically pointed out how to "rub hands" ((1) palm to palm and (2) palm to backs) and how to "rub fingers" ((1) thumbs, (2) interlocking, (3) backs of fingers to palm, and (4) tips to palm). Only these few words are needed because learners are watching the instructor perform the job of hand washing as they point out these important steps and key points. By practicing the proper procedure while remembering these critical factors, they would be able to perform the job correctly each time they did it from then on out.

Once this was accomplished, they then felt that the total time of the hand washing, the time it takes to sing "Happy Birthday," was really not significant any more. The original idea of 15 seconds was an attempt to make sure that proper time was taken to get to every part of the hands and fingers. But there was no guarantee of this happening, even if you sang the song, especially when people were never made aware of the proper technique, or if they felt they could shortcut the process. Knowing the proper procedure would ensure a correct practice every time without specifying a minimum time to wash.

The team got confirmation very soon that the procedure they were teaching was indeed the correct practice. One of the patient care technicians who had been trained early in the pilot just happened to be taking phlebotomy training, where she was learning how to draw blood, so that she could do some extra work in the lab. In the class, the first thing the instructor did was to show the importance of clean hands by having everyone put fluorescent gel on their hands, wash them, and then see what gel was still left, usually in the cracks of the fingers or along the edges of the fingernails, under a black light. The PCT who was trained how to wash her hands with JI was the only one in the class to pass the test. In fact, the instructor said he had never seen anyone get all of the fluorescent gel off and asked her to do it again, thinking it was a fluke. When her hands came back completely clean again, they spent the rest of the session having the PCT teach everyone in the class how to wash their hands properly.

There were many compelling reasons for the VMMC team to challenge hand hygiene as an appropriate task to teach hospital personnel. Since it was such a universal and visible function of the entire facility, success here could propel the use of TWI as a standard practice for the training of jobs throughout the hospital.

Training Rollout

Now that they had a good process to teach, the next challenge was to roll out the training to eight nursing units—467 RNs and PCTs—over a nine-week period (see Figure 9.4). They were determined to teach each of these people individually using the JI four-step method on three jobs: the two hand hygiene jobs (the job of washing hands with gel was basically the same as with soap and water but without the wetting, rinsing, and drying of the hands) and hourly rounding (described in the next section of this chapter). Moreover, they also planned to do follow-up checks with each person trained, up to five meetings, depending on the person, to make sure he or she was using the methods taught. With the typical hectic nature of a hospital ward, where nurses and other staff take care of patients on an ongoing basis, they felt that it would be difficult to pull people off the floor for training. And they didn't want to do the training as overtime because of the expense and because the nurses, who worked twelve-hour shifts, would not be focused on training at the end of a long day. They also thought that since training was part of their required work, it ought to be completed on the job, not after already putting in a full shift. The task seemed daunting, or, as Martha described it: "Of course, it cannot be done. We are, therefore, going to do it."

They started looking for downtime during the day that would be best for doing training. Looking at it the other way around, they tried to identify when there were flurries of work time each day when training would not be practical. Seven to ten in the morning was just one of these

Job Instruction Training Timetable—Healthcare

Job name:
1. Hand hygiene—soap and water
2. Hand hygiene—gel
3. Hourly rounding

Departments: 17, 16, 15, 14, 10, 9, 8, 7

Date: Summer 2009

	July 26–August 1	August 2–8	August 9–15	August 16–22	August 23–29	August 30–September 5	September 6–12	September 13–19	September 20–26	Changes in Schedule
Level 9	×	×	×							
Level 14	×	×	×							
Level 8			×	×	×					
Level 7			×	×	×					
Level 17					×	×	×			
Level 16					×	×	×			
Level 15							×	×	×	
Level 10							×	×	×	
Turnover										
Work performance										

Figure 9.4 Training timetable for nine-week rollout. (Reprinted with permission of Virginia Mason Medical Center, Seattle, WA.)

times. This is when the shift change occurs, so nurses coming on shift are busy transitioning care from one caregiver to the next, rounding on patients, and prioritizing their duties. Patients are also waking up and need to go to the bathroom, be served breakfast, get their medications, and so on. Also, this is typically the time when deliveries are being made so there is a lot of commotion sorting through incoming materials and getting it arranged and put away. They found, then, that 10 a.m. to 2 p.m. was a good time, as well as 5 to 9 p.m. in the later shift. So they blocked off four-hour chunks of time for training and made a daily schedule for when members of the JI-trained team would visit the various units (see Figure 9.5).

Even with this schedule, they still had to maintain a high level of flexibility. Within each two-hour spot they would plan to train four people, one at a time, in all three jobs. A week before going into any area, they would go over the schedule with the charge nurse, being respectful of their unique situation and looking for bits of downtime. They would try to preplan who would be trained, but if they found that the area was slammed the day of the training and the unit did not have individuals available to train, they would move to another area to continue the training. The rule was to always move backwards on the schedule and do follow-ups or pick up people who had missed the training in previous sessions. They wanted to maintain good communication and support with the units, so they tried not to jump ahead on the schedule to areas before they were expecting to be trained. The key was to partner with the areas so they didn't feel like this was "*us* coming to do *this* to *them*."

This week's focus units: 7, 8, 16, and 17						
August						
	Sunday	*Monday*	*Tuesday*	*Wednesday*	*Thursday*	*Friday*	*Saturday*
	23	*24*	*25*	*26*	*27*	*28*	*29*
12:00 a.m.							Joanie C.
1:00 a.m.							
2:00 a.m.							
3:00 a.m.							
4:00 a.m.							
5:00 a.m.							
6:00 a.m.							
7:00 a.m.							
8:00 a.m.							
9:00 a.m.							
10:00 a.m.		Martha Purrier	Martha Purrier		Janine W.		
11:00 a.m.							
12:00 p.m.				Charleen T.			
1:00 p.m.							
2:00 p.m.			Charleen T.				
3:00 p.m.							
4:00 p.m.		Janine W.		Arni V.			Alenka Rudolph
5:00 p.m.					Martha Purrier	Debbie K.	
6:00 p.m.							
7:00 p.m.							
8:00 p.m.							
9:00 p.m.						Joanie C.	
10:00 p.m.							
11:00 p.m.							

Figure 9.5 TWI training schedule. (Reprinted with permission of Virginia Mason Medical Center, Seattle, WA.)

Before they began, Martha looked at the overall schedule and made a running count of how many people needed to be trained each day in order to meet their deadline. Preparing a standard takt time calculation sheet, she figured they needed to stay on a pace of seven people trained per day. She continued doing the calculation on a weekly basis as the training moved forward, adjusting the daily total to be trained—sometimes the total would go up to as many as nine people per day. She worked with her team to stay on track even as they struggled with the daily challenges of getting people into the training.

The trainers were instructed to "resist the temptation to group train." When going to a unit, they would first check the training timetable for that area (see Figure 9.6) and deliver the training to people on the list one at a time. It was a common practice, especially for nontechnical tasks, to batch train people and send them directly out to the floors. But the results were always spotty, at best. The JI technique taught them the need to train "*a* person how to quickly remember to do *a* job correctly, safely, and conscientiously," meaning that this was one-on-one training. This was a huge commitment, but Martha was confident that the results would prove it to be a smart investment.

Nevertheless, for the small group of initial JI trainers, getting through 467 training sessions one at a time was a heavy lift. And to get them all done in nine weeks meant Martha had to encourage them and maintain high enthusiasm. A July 30, 2009, memo to her training team exemplifies this effort:

ITEM 2. Apparently, there is a limit to how many consecutive staff one is able to train before one can't remember their own name. If this happens to you, fear not! Your brain will remember the jobs perfectly ... no need for you to be present. Also, you will be wearing a name tag, and a nice member of our staff will assist you.

When a training session was completed, they would write the date it took place in the appropriate box of the timetable. They would also conduct follow-ups where they viewed the jobs being done in the flow of actual work. These could be conducted "as early as the same day as the training as long as the staff member has had a chance to get some real-time practice." During these follow-up visits they would not only watch them do the job, but also ask them how the training had gone, if they had had a chance to try out doing the job, how things were going with the job itself, and if they felt there was anything missing. There were five slots allotted for follow-ups on the timetable (see Figure 9.6), but that didn't mean they had to do all five. The instructions to the trainers were to continue conducting follow-ups until they felt the person had mastered the technique and successfully integrated it into his or her daily routine. If they felt the person needed more follow-ups, they would indicate this. Otherwise, they would write "done" next to the date.

They learned one thing at the outset when they began doing the training in front of the patients. Although patients at VMMC are oftentimes invited to participate in process improvement efforts, and usually do so with great enthusiasm, in this case they became very concerned while watching a staff person learn to wash their hands. They could not grasp that the learners already knew how to

Job Instruction Training Timetable—Healthcare

Name: Abolafya
Group: PCT
Department: ACE
Date: June–July 2009

Unit training minutes left: 1,500

Name	Training Minutes Left	Breakdown Number	Hand Hygiene Soap/Water						Hand Hygiene Gel						Hourly Rounding					
			Training	F/U #1	F/U #2	F/U #3	F/U #4	F/U #5	Training	F/U #1	F/U #2	F/U #3	F/U #4	F/U #5	Training	F/U #1	F/U #2	F/U #3	F/U #4	F/U #5
R., Brittany																				
M., Tatyana																				
S., Adrea																				
G., Ashley																				
B., Choky																				
F., Erlicita																				
A., Tshayinesh																				
D., Almez																				
G., Michelle																				
L., Iida																				

P., Greg															
F., Christi															
B., Mary															
M., Mary															
B., Marta															
J., Dawda															
P., Margaret															
E., David															
T., Alm															
C., Soleda															
C., Loma															
A., Amran															
A., Kiros															
W., Leah															
J., Annalyn															
Changes in schedule															

Figure 9.6 Detailed training timetable. (Reprinted with permission of Virginia Mason Medical Center, Seattle, WA.)

wash their hands and were just practicing the method of instruction. Patients began complaining that they did not want a nurse caring for them who was so inexperienced that she did not know how to wash her hands. So the team immediately made the adjustment to have all the training done in an empty room, and locating this space became the first task of the trainers when they went to the floors to teach. As the training started making progress, though, these very same patients began noticing the improved technique and started making comments about staff washing their hands "like they were going into surgery!" The effect was so large and immediate that Charleen, already a big proponent of TWI, began assisting with the JI follow-up checks to make sure nurses were using the proper hand washing technique.

One other benefit of giving the training to the patient care technicians who normally would not have received this level of intensive training with this much detail was providing the reasons why. Until TWI, few people had taken the time to explain to them the reasons they should do their jobs in certain ways. This more respectful way of training via TWI, providing the reasons why, created greater motivation and a more pleasant working environment. During the training process the PCTs asked many questions, for example, on other aspects of infection control, that never came up before, and their overall level of expertise grew dramatically.

The initial training rollout was completed on schedule in nine weeks. But before we go over the results, we need to look at the third job that was taught in this pilot, along with the hand washing and hand cleansing: hourly rounding.

Hourly Rounding: A Solution to a Universal Problem

One of the most persistent and costly hospital problems, in terms of actual dollars spent and amount of patient discomfort, is inpatient falls. This is a "tough nut to crack," as Martha put it, and all hospitals are trying to figure out ways to solve this problem. Costs involved with injuries incurred while in the hospital due to falls can be phenomenal, and the safety of patients would be greatly enhanced if a way could be found to reduce them.

While traditional approaches have been to look at the patients' physical and mental state in relation to the falls—weakened bodily conditions, lightheadedness, tied to medical equipment, etc.—and to try and mitigate these conditions in order to prevent falls, the VMMC group looked at the problem from a different root cause. For the entire year of 2008, they took data on where the falls occurred and what the patients were doing when they fell. The largest situation, 32% of total falls, occurred when patients were going to the bathroom. Of the eighty-seven people who fell while toileting, 57% of these occurred as they ambulated to and from the bathroom and 43% of these patients fell off the toilet. There were an additional twenty-three patients who were also found on the floor, and though the causes of the falls were not known, they may have been toileting related.

Nurses and hospital staff are always ready to assist patients with going to the bathroom. But patients are reluctant to ask for help or overestimate their ability to self-ambulate. Martha and her JI group felt they could tackle this problem by training staff to better approach the patients in a way that would make it easier for them to acknowledge this need and get help before a fall took place. But this kind of skill was different from the more routine jobs, like washing your hands or collecting a specimen, where there are actual hand motions and physical techniques. They wondered if the Job Instruction method could even be used here.

The inpatient leadership team felt that falls could be reduced by having nurses make hourly rounds of the patients where they would ask them if they had to go to the bathroom and offer help if needed. Here, they would have to appeal to the nurses' sense of being care-givers—to their "inner nurse" and the real reasons they came to this profession in the first place. They would have to find a way to create a sense that, as a care-giver, "it ain't no big thing to need a little help now and then." But they faced several obstacles. Patients typically were embarrassed to ask for help, even when offered, for this most private of needs. The patients reported that they "did not want to bother the nursing staff, who are already so busy."

Getting Patients to Ask for Help

First, they looked at how nurses approached patients and what they could say in offering help that would get an honest response. They felt that asking, "How are you?" was such a colloquial question that the answer would almost always be a simple, "I'm fine," no matter how poorly the patient might actually be feeling. They wanted a more therapeutic, clinical opening phrase, and many in the group felt the key word should be *pain*, as that would indicate something specific about the patient's condition: "Are you in any pain?" But not all areas of the hospital had to deal with pain, and most people might not consider having to go to the bathroom a kind of pain. At that, someone threw out the word *comfort* as a way of greeting the patient; an opening line that conveyed concern for their well-being was: "Are you comfortable?"

They weren't sure if *comfort* was the right word for the task so, in order to come to a stronger conclusion, the team was tasked to quiz a whole unit of about twenty-five patients on what they thought about the word *comfort* and to write down what they heard. To their big surprise, they got an avalanche of responses, ranging from going to the bathroom to how to order lunch to one person who said, "The social worker told me she would be back, but that was yesterday and I'm getting nervous about going home." The word *comfort* brought out so much more about what was really going on with the patients that they became convinced this was the approach they were seeking. The second important step to the breakdown, then, after the first step of knocking and announcing your entry, became: "Ask about comfort" (see Figure 9.7).

Next, when it came to actually offering them assistance to the toilet, they felt that this had to be done tactfully to respect the patient and maintain his or her dignity. In this way, they could mitigate the embarrassment around needing help to go to the bathroom. So the group looked at why patients were hesitant to ask for

Version 3.0
07.07.09

Job Breakdown Sheet—Healthcare

Task: Hourly rounding

Supplies: (Patient room)

Instruments and equipment: Bed

Number of instructions before job demonstration:

1. HH before and after this job
2. How this differs from being in room often
3. Three script choices for toileting

Important Steps	Key Points	Reasons
A logical segment of the operation when something happens to advance the work	Anything in a step that might: 1. Make or break the job 2. Injure the worker 3. Make the work easier to do, i.e., knack, trick, special timing, bit of special information	Reasons for the key points
1. Enter	1. Knock 2. Announce	
2. Ask about comfort		
3. Offer toileting	1. Assess mobility status 2. Document elimination	Get help/assistant
4. Items within reach	1. Call light 2. Table	
5. Check bed setting	1. Low/locked 2. Alarm 3. Surface power—heel zone	Reduce skin breakdown and falls
6. Leave	1. Ask if anything else 2. Tell planned return 3. Initial grid	

Figure 9.7 Breakdown sheet for hourly rounding. (Reprinted with permission of Virginia Mason Medical Center, Seattle, WA.)

help, and they found they could categorize the reasons into three types:

1. **Independent people** who find themselves unexpectedly in the hospital with limited mobility and find it mortifying to need and ask for help to go to the bathroom
2. **Chronically ill people**, such as patients with cancer, who consider it a setback in their recovery to find themselves, again, in the hospital, and thus deny the fact that they need help getting around in an attempt to prove that they are getting better
3. **People with total loss of faculties**, such as those with dementia, who are not in control of their bodily functions and thus not even aware of their need

They felt that they would need specific and different strategies to address each of these groups, and so the key point here was to "assess mobility status" and take the right tact. They came up with phrases that they could use in each case and practiced giving these lines during the training sessions. For the first group of independent people who felt shame in needing help to the bathroom, they would ask, "I want to measure your intake and output, so let me know when you go." For the second group of chronically ill patients who needed help but were reluctant to ask, they would say, "It's been a few hours since you last went. Would you like to go now? I have time." And for the final group of people who didn't know when they needed to go, they would say, "It's been two hours since you last used the bathroom, so let's go now." In this way, by determining the three types of mobility status, they were able to maximize the effect of the rounding service.

One more line they found to be quite effective, after asking if the person needed help, was, "I can help you. I have time." Most patients tend to be very cognizant of the busy nature of the nurses' jobs and are afraid that their own personal needs are interrupting that routine. By letting the patients know that it is a good time to be asking for help, this further reduced the barriers to seeking out and getting that help.

Once the toileting is completed and documented, the next two steps in the hourly rounding process are to check "items within reach," especially the call light so they can call for help if needed, and "check bed setting" so that the bed is in the proper position to minimize the effects of a fall if they do try to get up. These two steps provide some backup against falls during the time between the rounding visits. While making this breakdown, they also kept in mind Lean concepts such as flow in the work. Initially, the breakdown had checking the bed settings, which are located at the bottom of the bed, before tidying up the bedside table. However, as they walked through the job, they found that the flow of the work through the room would be shorter, with no backtracking, if they did the bed settings last, on the way to the door and out of the room. They were able, in this way, to teach the job in the most efficient way without any waste of time or walking.

Finally, the key point to the sixth and final important step, "leave," after asking if there is anything else they need, is to tell the patient when you plan to return—in an hour, in fifteen minutes with the next medication, in half an hour with a meal, etc. This innovation has proven extremely effective because patients can judge whether they should put on the call light if they need to go or just wait until the nurse returns. As it turns out, patients agonize over whether they should call a busy nurse, even if they are in great pain. If they know when the nurse is going to be back, they can make a better choice of whether to wait or go ahead to push the call button.

Martha experienced the power of this method when she was following up on one nurse who wanted to show her skills at rounding. She was the charge nurse, and they visited a man who was not her regular patient, so they did not know each other. The young nurse, sweet and patient, asked him, "Are you comfortable?" to which the older gentleman just growled at her. "It's been a while since I've gone to the bathroom," he continued. "OK, let's go, I have time to help you." He reluctantly went along with her, and after they returned, the nurse checked the bed settings after making sure everything was in reach and tidy on the bedside table. "So, is there anything else you need?" she asked. He scowled. "Well, I'll be back in an hour with your lunch." At this point the man, who had been grouchy and rude the entire time, turned to her and said in a low voice, "Well, it'll be nice to see you again." They found out what his real problem was—he was lonely. And that was information they could act on.

In addition to preventing falls by helping people get to the toilet, this method, perfected through many trial runs and modifications, was also able to provide the staff with much needed and useful information from the patients. What is more, it improved the customer service and the overall image of the hospital to its patients. And this soft skill, it turns out, could most certainly be taught and practiced using the JI method.

"We're Too Busy to Do This!"

When the JI team began teaching the hourly rounding routine to the staff, the first reaction they got was, "We're too busy to do this!" So this job instruction needed a great deal of tact and skill to teach effectively. The trainers were teaching the job at the same time as they taught the two hand hygiene elements, but while those parts were fairly straightforward and RNs and PCTs could immediately understand and accept the need for correct hand washing, they felt that the rounding procedure, something new that was to be performed hourly, would unduly impose on their already hectic work schedules. Having experienced the teaching of the rounding job while fine-tuning the breakdown and getting the procedure working the way they wanted, the team also perfected the way they would train it in order to overcome this resistance.

Instead of simply telling learners when to perform the rounding work, they devised a process of self-discovery where they would let the skeptical nurses and nursing assistants find out for themselves when they could insert the routine into their work. They would ask them, "What are you doing already, like taking vital signs, delivering food, giving medications, and so on? How would you fit this into that routine?" Then they would let them actually practice it, walking through the routines in the empty room they were using for training to see what it would actually look like to perform the rounding steps as they did the work they were already doing. They would practice under different scenarios, thinking it through thoroughly, and that is when "the light bulbs start going off."

For example, if a trainee was attempting to do a certain task, like checking vital signs, while rounding, he or she might say, "I can put it between steps 4 and 5."

"Try it out," the trainer would advise.

"Well, it doesn't work real well there, does it?"

"What else could you do?"

"How about putting it between steps 2 and 3?"

They would try that out and find that, in fact, they could perform the rounding tasks without adding significant time to the work they were already doing. But they had to discover that for themselves to be convinced that the procedure could be performed in an effective and efficient way. This experimentation was done during the training of the job at the end of Step 3 of the JI four-step method, the Try-out Performance. Once the person learned the standard method, the trainer would ask him or her to do it one or two more times, integrating jobs he or she already did on a regular basis. The amended breakdown would look something like this:

1. Enter
   a. Knock
   b. Announce
2. Ask about comfort
3. Obtain vital signs (the job already being performed)

4. Offer toileting
   a. Assess mobility
   b. Document
5. Items within reach
6. Check bed setting
7. Leave

Right from the start of the training, the follow-ups the team performed began showing that they were getting strong reliability, in other words, the procedure was being performed regularly. One morning the team came in and heard that there had been a fall the night before on a unit they had trained. They were very concerned and went to investigate right away. Since the last step of the rounding process includes initialing the grid to show that the event has taken place, they went to the log to see if the rounding had been completed, and sure enough, there were two hours in the middle of the night where the patient had not been visited. That is precisely when the fall occurred. The team then began work on how they would arrange tasks so that hourly rounding would be done reliably, every single hour. They rearranged the sequence of work and inserted signaling so that they could achieve a higher degree of reliability.

Results of the Initial Rollout

The first signs that the training pilot was having a positive impact came from comments the staff began hearing from the patients themselves, who, as the hospital's "customers," are always auditing the performance of the care they receive at VMMC. Ellen Noel, a Med-Surg clinical nurse specialist, made this observation:

> Recently I entered a patient's room on level 10. From her bed, this patient watched me wash my hands. The patient remarked, "That is so interesting! *Everybody* coming in here washes their hands the exact same way. I've never seen anything like it!"

Even outside of the hospital, people were noticing. One of the nurses trained in hand washing was using the restroom before going into a theater to see a movie and, while washing her hands, another patron noticed her method and commented that she must work for Virginia Mason because "that's the way everyone washes their hands there." Then the person asked if she would show her how to do it correctly and, before she knew it, several other women in the restroom were observing and learning how to wash their hands correctly.

With the rounding procedure, as well, patients were noticing the standardization and reliability of the service they were getting. The following conversation was overheard between a patient being discharged and his nurse in the telemetry unit:

> You know ... you all must go through some kind of special training because *everyone* asked me if I was comfortable, offered the bathroom, made sure that I had my call light and phone, and then asked if there was anything else I needed. I've never seen such great customer service while in a hospital.

Notice here that the patient was able to recite the job procedure, completely and in proper sequence, simply by watching it being performed exactly that way each time. This demonstrates that the procedure has, in fact, become the *standard* way it is done by everyone on the floor. Notice also that, from the patient's perspective, this procedure was all about great customer service and not fall prevention. So the hospital was able to obtain a significant benefit in patient satisfaction and reduced falls.

The steadiness and reliability of the processes could be locked in through the JI training process—through learning properly how to do the jobs—and this proved more effective than simply telling people how important it is to do these tasks. Rowena Ponischil, director of levels 7 and 8, stated it this way:

> For a long time now, I've taught my staff that the majority of patient falls occur during the toileting process. Knowing, however, wasn't enough to hardwire actions to prevent patient falls. TWI provides the hardwiring and rigor … toileting is planned for and built into my staff's workflow. It's really made a difference on Level 8.

Hand Washing Pilot Created "Pull"

Earlier in this chapter we talked about the benchmark figure used to check the percentage of time healthcare workers washed their hands when they should before interacting with patients. We noted that worldwide that number was as low as 40%, and even at VMMC the 2009 figure was 83.5%. In the areas where the JI pilot was run, eight nursing units for a total of 467 RNs and

PCTs, reliability of hand hygiene went to above 98%. For the rounding procedure, reliability was also measured at over 98%, and where rounding was implemented, patient satisfaction scores were up 5 to 10%.

Besides these initial feedback statistics, the most immediate result of the pilot was the flood of requests from other areas of the hospital to do the hand hygiene training. This pull from other departments meant very clearly that they all saw the value and need for the same training within their own areas. Departments that were scheduled to continue with the hand washing training included:

- Transporters
- Outpatient clinics
- Hospital nursing units not part of the initial trial (ED, RHU, CCU)
- Pharmacy
- Bailey-Boushay House (a nursing home run by VMMC)
- Sterile processing
- GME (graduate medical education) and safety curriculum team
- MD Section Heads
- Surgery Section

In this way, the trainers who did the pilot never stopped training but continued rolling out the program into the pull from these other areas. All transporters throughout the hospital were immediately trained, and they began going into the clinics directly thereafter.

Even as the hand washing pilot was still being conducted, many areas had already begun seeing the need for and planning the use of JI for other tasks. In anticipation of this surge in demand for good job instruction technique, in September 2009, just as the pilot was winding down, Martha and one other member of the KPO staff, Alenka Rudolph, took the JI train-the-trainer program to become facilitators of the JI ten-hour course. Once the pilot finished, Martha and Alenka began teaching JI ten-hour sessions, only admitting carefully selected individuals to take the training to be sure the program would continue growing in a strategic direction with continued momentum.

One good example of that growth was in the operating room, where they had identified an immediate need for JI training: specimen labeling. Different from taking a blood or urine sample, where a test can easily be redone if in doubt, when a specimen is taken from a person's body in the OR, there is an "extreme chain of command" where the staff member responsible for the specimen must not lose sight of it until it is properly labeled. In other words, it should never leave their hands. In spite of the protocol in place, though, they had experienced some near misses where specimens were mislabeled and they wanted to eliminate the situation. Though these types of incidents were corrected without having to go back into the OR to extract another specimen, the near misses pointed out the need for a better process.

Members of the OR team went into the JI class specifically with the intention of redesigning this process. While doing a breakdown for the job, in preparation for a practice demonstration of the JI method using an actual job from the trainee's worksite that each person attending the class must perform, they asked the question, "Why is it seventeen steps?" Using the JI concept of important steps and key points, they were able to clear out the clutter and confusion in their current procedure and create a clear and effective teaching process that would ensure specimens were handled correctly each and every time they were labeled. As Martha pointed out to them, "When you design a process well, you do find the best way, just like the path that water takes when it flows downhill." But they first needed the JI model to show them that path.

Another example of how they saw the use of the JI method for problems they were trying to address was in the fire safety procedures at Bailey-Boushay House, a nursing home facility run by VMMC. In fact, fires in nursing homes overall are a big problem, and Bailey-Boushay had had three fires in three years, an unacceptable number, mainly due to a high population of smokers who did not always follow the rules around their habit. Bailey-Boushay followed national fire safety guidelines by teaching staff how to react to a fire using the acronym RACE: *remove* the patient from the area of risk, *activate* fire alarm, *call* for help, and *evacuate* if directed to do so. Even though people could remember the acronym, they didn't always remember what they were supposed to do.

The executive director of Bailey-Boushay, Brian Knowles, went to the JI class and used this fire procedure as his demonstration job in the training. He reframed the procedure into important steps and key points, the same steps as outlined by RACE, but without the acronym. After practicing teaching the procedure in the JI class, he went back to the nursing home and began training it to his staff. In subsequent fire drills, Bailey-Boushay staff not only remembered how to do the procedure, but were actually able to perform the actions prescribed proficiently and in a timely manner.

Martha knew that all of these priority jobs were problems that each area had already recognized, and their staff knew what the right response needed to be in order to solve them. In other words, TWI didn't tell them what they had to do. However, in spite of this understanding of the need for good procedures and training, they didn't know how to frame the solution or how to "edit it correctly," as Martha puts it. TWI gave them a solid methodology they could follow in order to finish the job of developing true standard work that solved these pressing problems. "Just knowing you have to do it doesn't get me to do it," Martha explains, "but if I know how to do it, and why, then I'll follow the correct procedure every time."

Patient Falls Were Reduced

Sometime before TWI was introduced to VMMC, the effort to reduce falls using VMPS tools was put into practice and good results were obtained. This became an organization-wide initiative in 2007. The average number of patient falls per 1,000 patient-days was 3.33 in 2006, and they were able to reduce this total to 3.04 and 3.10 in 2007 and 2008, respectively. More significant than the total number of falls, however, is the number of falls with injury—the ones that really count. This figure, as well, was reduced from 1.04 in 2006 to 0.77 in 2007 and 0.99 in 2008.

As we have described in this chapter, the initial TWI training of ten JI trainers was done in May 2009, but the pilot training of the hourly rounding job was not put into full production until July 2009, when 467 RNs and

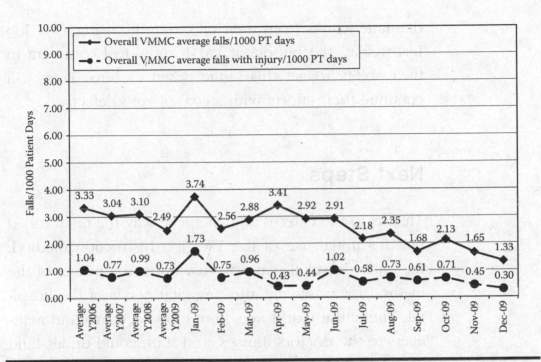

Figure 9.8 Falls and falls with injuries/1,000 patient days overall. (The above metrics are a snapshot of that time at VMMC. Results may differ at another time or under different conditions.) (Reprinted with permission of Virginia Mason Medical Center, Seattle, WA.)

PCTs were trained over a nine-week period ending in the middle of September. Looking at Figure 9.8, we can see the number of patient falls on a month-by-month basis drop off dramatically beginning in July and holds steady, if not continue to fall even lower. Even with just half a year of training results, the average for the entire year of 2009 went down to 2.49 patient falls per 1,000 patient-days and 0.73 patient falls with injuries. For just the six-month period since July, the totals would be 1.89 and .56, respectively.

Patient falls had always been considered inevitable at the hospital, one of those things that, in spite of their best efforts at prevention, would always be one of the "costs of doing business." However, inspired by these dramatic results from their pilot training using TWI Job Instruction, the people at VMMC now feel confident in their ability to get this figure down to zero. They will continue their efforts with a goal of zero defects.

Next Steps

Thanks to the careful and energetic way the team rolled out this initial use of the TWI Job Instruction method, the entire hospital was able to witness the power of the method for positively affecting vital needs of the hospital. The entire workforce, from top executives and managers to the doctors, nurses, and staff, could understand the method and advocate its use.

One area in which they feel they will get exceptional value out of a JI implementation is in telemetry, that area between critical and regular patient care where staff need to learn the use of a variety of monitoring equipment as they care for patients in a high state of need. This is a bottleneck in the VMMC organization because of the unique expertise needed to care for very ill patients. There is also high staff turnover, not because people are unhappy, but because they become very marketable once they are able to read the monitors and treat critical care patients. Nurses coming out of telemetry typically move up quickly to higher positions in clinical care, ICU, surgery, or even, in some cases, to becoming a flight nurse. So the need for good training here is constant, and the department has become a "feeder line" for the entire hospital.

The telemetry unit has embraced the fact that they are going to be a training point for nurses and staff moving up their career paths, and they have now embraced TWI as a steady means of providing this service to the organization. They ran with the initial hand hygiene pilot and, because of high morale in the department, got lots of questions, such as on the proper use of personal protective equipment (PPE), when doing the training and these questions were passed up the organizational chain. Management thought that everyone knew these things and were surprised they didn't. In response, they have now taken infection control procedures to new heights and are looking forward to using TWI to actualize these new standards, even as they continue training the new nurses who move into the department.

In addition, VMMC took a look at its graduate medical education (GME) program, which trains medical doctors starting out in their clinical practice education, and the clinical safety curriculum where reliability around high-risk procedures is critical. The hospital is working with the Institute for Health Care Improvement, which is considering what critical elements around safety need to be a part of everyone's medical training, including hand washing, which is now being included as part of training at the national level. They always knew that hand hygiene was a critical topic to include, but thanks to the work of the JI pilot, they now had a good way to teach it. They feel that TWI must be included in this part of the curriculum, in particular in the simulation labs they have set up as part of the program. There are a variety of topics that are covered in these simulation labs, but a good 30% of this training is now going to be straight-out TWI instruction.

The hospital also feels that it has a large responsibility in training the next generation of nurses and VMMC staff help teach at the many nursing schools in the Seattle area. With the success of the TWI pilot, they plan to show the JI technique at these nursing schools and require nurses to learn it if they want to do their clinical training at VMMC. Just as Toyota helps its upstream suppliers by teaching them techniques that develop their own quality delivery, VMMC feels it can help nursing schools better prepare future nurses who will eventually work for VMMC or other hospitals.

Donna Smith, the medical director of the hospital who is still a practicing clinical physician in pediatrics, participated in one of the first ten-hour JI sessions put on by Martha and Alenka. While the plan was to have participants in the classes be practitioners who would actually train jobs, in order to get understanding and support from the top brass, they had select administrators participate—but never more than three in the ten-person classes. As her practice job for the class Dr. Smith obtained a throat culture where you reach into the back of the throat with a swab and, as a key point, "don't touch tip until you reach tonsils," so that the patient, in particular a child, doesn't gag on the swab. The breakdown and instruction practice went extremely well, and the medical director felt that, with this training, you could get a nurse's assistant to be able to do this job instead of a doctor or a nurse. "There is no reason for me to do it," she proclaimed, "and no reason for a nurse to do it either."

Based on epiphanies such as this, they feel now that the JI method can be used to train nurse's assistants to take on some of the tasks commonly done by the nurses themselves. With the chronic shortage of nurses in every aspect of healthcare, this prospect is one of the most appealing results of the TWI introduction. Of course, each state has its own laws surrounding what jobs must be done by licensed medical professionals, and the scope of practice for assistant personnel may be limited. However, most regulations allow for jobs to be done by assistants if they are (1) under the supervision of a nurse and (2) if they have received proper training in the task. Some of the jobs that might fall into this category are:

- Connecting an EKG (hooking up the electrodes)
- Taking vital signs
- Discontinuing an IV heplock
- Taking out an IV
- Removing a Foley catheter

These are jobs that are not currently part of the training curriculum for nurse's assistants but ones that can be trained and managed by the nurses.

Another vital area where they feel JI can create large gains is in presurgery procedures and preparation. Hospitals have long practiced things like the procedural pause, where they take a moment just before surgery to make sure, oftentimes with the patient's participation before they go under anesthesia, that they have the right patient for the right procedure with the right equipment and expertise in the room. These procedures would only be enhanced by having everyone trained properly using JI.

In addition, VMMC leadership feels that a huge opportunity exists in the presurgery area of sterile processing, where all instruments and equipment in the operating room must be cleaned, sterilized, and rekitted for next use. This is the one area of the hospital that actually operates similar to a production line in a factory with, quite literally, a moving line, peak periods of output, and operators with well-defined tasks. The staff and employees here have tried valiantly to improve these process, attempting to standardize work and better sequence the flow of the processes. They have approached improvement with great dedication, but they struggle with stability, and therefore there is no visibility to find the vital areas that need attention. With JI, leadership feels, they can reach that needed stability and better reap the benefits of the VMPS tools they are applying.

Another area of great potential is in the pharmacy, where the compounding of medications is a critical function. In other words, while many medications are prepackaged and prepared for delivery, hospital pharmacy staff still mix ingredients to prepare medications in specific dosages for individual patients. Needless to say, this must be done carefully and with no mistakes. For example, in chemotherapy, drugs are compounded based on body height and weight, and it is easy to transpose dosages when mixing the formula. If the correct dosage for a patient was, say, 60 mg/m^2 of Adriamycin and 600 mg/m^2 of Cytoxan for a particular patient, an error of mixing 600 mg/m^2 of Adriamycin and 60 mg/m^2 of Cytoxan would create catastrophic results.

To guard against these kinds of errors, a system of checks and balances has been set up to capture mistakes, and herein lies a great deal of waste and redundancy. But the safety aspect requires this redundancy.

In fact, at VMMC they were finding corrections 60% of the time. With more consistency in the work, one wonders, could we create more stability in this critical area and throttle back on some of the redundancy in these checks?

Conclusion

The possibilities for substantial improvement in healthcare using TWI Job Instruction seem almost endless. As they move forward, the staff at Virginia Mason Medical Center will continue to reap these benefits of good job instruction as they continue to apply the training purposefully and strategically. What is more, their many years of continuing effort implementing Lean under their VMPS system will garner even better results because of the strong foundation of good leadership and work skills they are planting with TWI.

IV

TWI IMPLEMENTATION

Chapter 10

Applying TWI to Your Organization

> The 10-hour sessions of JI, JM and JR are so basic
> that their importance is not subject to debate.
> Completion of these sessions, however, is only the
> beginning. The continued use of these skills over
> the weeks, months and years is the only way to
> get results.[*]
>
> **—C. R. Dooley, July 1944**

When the TWI Service was first created by the Council
for National Defense in 1940, it had twenty-two district
offices around the country. These offices provided consulting on in-plant problems with a staff of experienced
industry people and vocational education instructors
on loan from their organizations for the duration of the
war.[†] The first challenge presented to this group was
to address a critical shortage of lens grinders needed to
produce precision lenses for wartime use in the bomb
sights bombardiers used to hit their targets. Studying the

---

[*] C. R. Dooley, *How to Get Continuing Results in a Plant from Training Within
Industry Program*, Training Within Industry Service Statement Policy, July 1944.

[†] Alan G. Robinson and Dean M. Schroeder, Training, Continuous Improvement,
and Human Relations: The U.S. TWI Programs and the Japanese Management
Style, *California Management Review*, Vol. 35, 37, 1993.

situation, the TWI specialists found that the job of lens grinding appeared to require the mastery of twenty jobs, not all of which were skilled. The shortage of lens grinders to meet the wartime production justified the training of relatively unskilled workers who had no reservations about doing the simpler tasks. TWI staff also improved the training for new lens grinders, reducing that time from five years to two months.

The TWI Service used this lens grinding success story to showcase the power of proper training, and this led the service to concentrate solely on training rather than consulting, which was their initial thrust, to improve quality and productivity nationwide. They then went immediately to work on creating the J programs, beginning with Job Instruction that, after numerous test runs, was put into operation as a national program in October 1941.

Success with the lens grinding project led to the inspired principle that continues to this day to be the key to a successful rollout of the Training Within Industry programs: the multiplier effect. As Walter Dietz, one of the TWI founders, explained in his 1970 book, *Learning by Doing*, the foundation for the multiplier effect is to "develop a standard method and then train people who will train other people who will train groups of people to use the method." For the multiplier effect to work, the J courses had to be designed as a standardized method that could be successfully taught in all situations in a variety of industries. Once taught, stringent quality control would be needed to sustain the multiplier effect of the training. This combination of standardized training and follow-through was established with JI and then successfully repeated for JM and JR.*

To illustrate how the multiplier effect worked, 1,305,570 supervisors had been certified to teach the JI ten-hour sessions. These people, in turn, trained over 10 million workers, one-sixth of the entire workforce of 64 million when TWI ceased operations in 1945.† However, this multiplier effect was triggered only when the training of the J programs was properly delivered by a qualified instructor who had been trained to deliver the programs following the trainer manuals that had been developed by the TWI Service.‡ This strict adherence to standard delivery locked in the quality of the content that spread rapidly through the workforce by those teaching it.

A Plan for Continuing Results

As each program was introduced, the large volume of people in need of the training provided the TWI Service with feedback from trainers and from companies throughout the United States. That led C. R. Dooley, head of the service, to issue a national policy on how to get continuing results in a plant from Training Within Industry in June 1944, near the end of the war, to all Training Within Industry staff members. "This plan for getting 'continuing use' of TWI programs," he wrote, "is

---

* Ibid., p. 40.
† Ibid, p. 38.
‡ Patrick Graupp and Robert J. Wrona, *The TWI Workbook*, Productivity Press, New York, 2002.

the out-growth of two years of practical experimentation and experience. You will recognize many features of the plan as your own. It is another splendid example of the combined efforts of many TWI people."

Dooley's statement of policy separated the implementation of TWI into two phases:*

Phase 1: *Basic training.* The initial introduction and follow-through efforts with JI, JM, and JR. Albany International, W. L. Gore, Virginia Mason Medical Center, and Currier Plastics are examples of organizations in different stages of phase 1—giving people the skills to improve in the ten-hour sessions of JI, JM, and JR.

Phase 2: *Continuing use.* All of the things companies do to see that supervisors make continued application of the principles given them in the JI, JM, and JR ten-hour sessions. The use of JM to drive the Donnelly Custom Manufacturing idea program is a good example of how one company has devised a way to keep its line organization focused on using TWI to continually improve.

Applying what is learned in the ten-hour sessions may differ between companies or even between plants within the same company. However, the same positive results will be generated when each plant conforms to a fundamental plan—something that was lacking at Nixon Gear after its initial TWI training. The planning process

---

* Dooley, *How to Get Continuing Results in a Plant from Training Within Industry Program.*

developed in 1944 for introducing TWI and for getting continuing use of JI, JM, and JR is a good place to start our review of how an effective TWI rollout can be properly managed:

Phase 1: A plan of action to suit each plant. Commitment from top management to:
1. Sponsor the program
2. Assume the function of a coordinator or to designate someone in that role
3. Approve a detailed plan for basic training and continuing use
4. Check results

Phase 2: Plan for getting continuous results from JI, JM, and JR:
1. Assign responsibility for getting continuing results.
    a. Each executive and supervisor must:
        i. Use the plan himself or herself
        ii. Provide assistance to those who report to him or her
        iii. Require results of those who report to him or her
2. Get adequate coverage.
    a. The ten-hour course is not just for first-line supervisors. Third-line and fourth-line management, along with people at all levels, should get the training to know what the program is about.
3. Provide for coaching of supervisors by either:
    a. Their own boss
    b. The person best able to influence them

4. Report results to management.
   a. The appropriate executives should be regularly informed as to results in order to have some continuing connection with the program and some incentive for continuing interest.
   b. The purpose of the report is to show the relative improvement accomplished by the use of JI, JM, or JR.
5. Give credit for results.
   a. Prompt and proper recognition by the appropriate executive is necessary to obtain continuing interest, and hence continuing use.
6. Provide for clearance of Job Methods proposals.
   a. A well-set-up procedure for handling proposals is a must.

When this plan was put out at the end of the TWI Service's life, the timing to blaze new trails could not have been worse. Manufacturers in 1944 were already shutting down the production of war materiel while ramping up the production of consumer goods now that an Allied victory was no longer in doubt. Management had come to view TWI as a wartime program, so taking the TWI training to another level to get continuing results from JI, JM, and JR was not on their list of priorities. TWI had also provided grassroots attention to how people were treated in the workplace, and this made management uncomfortable at this time in history.

After the war finally did end, TWI soon traveled to Japan, along with Juran and Deming, at which time the Japanese Labor Ministry set up a small TWI working group. This group sent one member abroad to a JIT Institute* directed by the International Labor Organization that had obtained copies of the TWI manuals. This person returned to Japan where he trained ten of its members as TWI institute conductors (master trainers), who trained approximately five hundred people in JI and seventy people in JM. However, this effort did not trigger the multiplier effect as was done in the United States during World War II.† Recognizing that the training had failed, the Japanese Labor Ministry then contracted with TWI, Inc. in 1951 for three master trainer institutes, one each for JI, JM, and JR. A specialist for each J program was selected to travel with them to Japan from the United States to deliver the TWI programs. A fourth specialist, an expert in plant installation of the TWI programs, accompanied the trainers. This person was charged with selling the value of the TWI programs to top management. According to the newly arrived TWI, Inc. specialists, the failed efforts that preceded them had lacked the rigid quality control and attention to detail in the delivery of the ten-hour programs to trigger the multiplier effect.‡

These specialists introduced TWI in Japan in the same manner they had learned worked best while training TWI throughout the war in the United States:§

---

* The Americans originally labeled the programs JIT for Job Instruction Training, JMT for Job Methods Training, and JRT for Job Relations Training. The Japanese later took out the common *training* label, and we use that convention today to indicate the methodologies themselves and not just the training aspect to the programs.

† Robinson and Schroeder, Training, Continuous Improvement, and Human Relations, pp. 40, 47.

‡ Ibid.

§ Ibid.

- They publicized their arrival and the J programs they were to install through the press, and government and military authorities.
- They helped to select trainees from a wide variety of industry groups.
- Unions were invited to participate to learn that TWI was to train people to become better supervisors, not to make laborers work harder.
- The TWI plant installation specialist worked with Japanese partners to sell the value of TWI programs to top management.
- JI, JM, and JR trainers delivered the standardized TWI training to groups of ten people.
- TWI, Inc. selected and prepared follow-through trainers, quality control specialists, installation specialists, and a core of institute conductors (TWI master trainers) to train instructors.

"When the TWI, Inc. specialists departed from Japan, they left behind them 35 certified Institute Conductors (Master Trainers), the beginning of a large multiplier effect which would extend to over one million Japanese managers and supervisors by 1966, and to many millions more by 1992."[*]

Lessons from the Past

The rollout of the TWI program in the United States and in Japan are the only examples of successful national rollouts of TWI from the past. The programs and what was learned by the application of these programs during

---

[*] Ibid., pp. 47 and 48.

WWII is extremely well documented, and there is no disputing the results. We also learned that the initial Japanese effort to implement TWI failed when the ten people trained by the International Labor Organization, using manuals and materials obtained from the ILO to train five hundred JI "trainers," were not able to trigger the multiplier effect in Japan. Albany International Monofilament Plant (Chapter 6) initially made the same mistake of using materials obtained from the archives in conjunction with exercises taken from *The TWI Workbook* without knowing how to apply JI, JM, and JR in the workplace. Both of these attempts failed because the people missed out on guidance from a certified trainer who is skilled in coaching people on how to apply what was taught in the ten-hour training classes.

Toyota TWI

Toyota was one of the first companies in Japan to embrace TWI for the very reason the program was created—to solve specific production problems involving people. According to Isao Kato:

> In 1950, during the near bankruptcy period, management and the union made a series of agreements. One agreement was for Toyota's management to respond to the union's request to create some form of supervisor development and training. The HR department of Toyota investigated existing programs and was introduced to TWI. Since it was an existing program and was receiving favorable reviews, it was evaluated and then adopted by Toyota.[*]

[*] TWI Influence on TPS & Kaizen, summary notes from Art Smalley interview with Mr. Isao Kato, February 8, 2006, www.artoflean.com.

Taiichi Ohno, the founder of TPS who would himself become a TWI-certified trainer, was very aware of the TWI programs from the start. Isao "Ike" Kato, who worked extensively developing training material for TPS under the direction of Mr. Ohno and other executives, became its top executive coordinator. It was interesting to learn from Art Smalley's 2006 telephone interview that Mr. Kato is known as the "father of standardized work and kaizen courses," and that he is also a master instructor of the TWI material. Mr. Kato cautions us in this interview that "you cannot separate people development from production system development if you want to succeed in the long run."

> TWI had significant influence on the development of our thinking and the way we structured supervisor training.... It did give us a vehicle to enhance supervisor skill sets and influenced the development of kaizen training courses, that is certain. It helps build capability into the organization at the supervisor level, which is very critical for TPS to succeed. TPS will not flourish if just staff and engineers are driving it from the side.... If people want to succeed with Lean or TPS, they have to emphasize people development and making leaders capable of delivering improvements. TWI is a great starting point even today and is a hidden strength of Toyota's production system.[*]

Getting Started on the Right Foot

If you have followed our TWI discussions this far, you have realized, several times over for sure, that we lean

---

[*] Ibid.

almost exclusively on the wisdom of the TWI founders to show us how to do TWI today. The fact of the matter is that what they taught us really works. What we are looking at in this chapter, then, is how to roll out the TWI programs so they become part of the everyday culture of an organization. This was successfully done during WWII in the United States and after the war in Japan, as we have described above. What is left for us to consider is how we can do it again today. The case studies in this book demonstrate how companies have taken up this challenge and wrestled with it to a successful conclusion. Here, let's sum up what we have learned from these endeavors, always mindful of the history of TWI that has guided us to this point.

Form a TWI Working Group Responsible to Lead the Way

Interest in TWI is typically initiated by one or two people who, one way or another, recognize TWI as the answer to pressing problems within their organization. For the company leaders to then select the right person to lead the implementation is a critical step to ensuring success. The initial vision Alan Gross of Currier Plastics (see Chapter 11) had for TWI is now shared by Dustin Dreese and Jennifer Dietter, who are working diligently with the Currier people, putting JI, JM, and JR to use in the workplace. Linda Hebish of Virginia Mason Medical Center discovered TWI at a TWI Institute workshop conducted at an Association for Manufacturing Excellence conference, but it took people like Martha Purrier and Alenka Rudolph to drive the initial pilot projects and then to lead the expanded use of TWI as certified trainers. The turn-around at Albany International initiated by Scott Curtis would not have occurred had it not been for Scott Laundry, Jamie Smith, and Jennifer Pickert, who helped to rally the workforce on his behalf. The same is true at Donnelly Custom Manufacturing where Ron Kirscht, Brad Andrist, and David Lamb partnered with Sam Wagner who introduced them to TWI. Mark Bechteler of Nixon Gear and Suzanne Smith of W. L. Gore, once they signed on for leading the TWI charge, literally "walked through the fire" in order to overcome obstacles that threatened to trip up their vision for TWI in their organizations.

These people are passionate about TWI because they see how the training changes the way people think about themselves, about others, and about their work. This passion rubs off on other employees as they too begin to see the value of TWI. For example, we arranged for representatives from a very large company with numerous plants and operations around the globe to visit Albany International, where Jamie Smith and Jennifer Pickert conducted a tour. The company's North American VP of manufacturing told us afterwards that he was not initially that interested in TWI but, after visiting AIMP, "I want my people to talk about their jobs with the same excitement that Jamie and Jennifer have for their jobs and the TWI training."

However, it takes more than two people to generate the impact TWI is having at the Albany International Homer plant. It takes a cross-functional working group as led by Scott Laundry and a supportive leader like Scott Curtis to move TWI from talk to action. Practicing the TWI skills is an activity that must take place across the entire organization; no one person alone can achieve this big a transformation. The initial TWI leaders must put together a team of people, a TWI work group that will work together for TWI to take a firm hold. Interestingly, it is not unusual for people to come forward and, without solicitation, volunteer to be on the TWI team after participating in the training because they want to be part of the transformation.

Once formed, this TWI work group must first attain a fundamental understanding of the TWI program before they try to sell it to the organization. Fortunately, there is a wealth of information about TWI on the Internet, in articles, and in books that are available to get a team started in the right direction.

1. For those interested in the history of the program, visit the Green Mountain Chapter of the Society of Manufacturing Engineers (SME) at http://chapters. smc.org/204 where you will find historical information related to Training Within Industry.
2. Numerous books and articles have been published about TWI and its connection with Lean and the Toyota Production System. Visit www.twi-institute. org for a list of recommended books and articles to start down that path.
3. Contact the TWI Institute to learn about companies in your industry that are using TWI that you can visit or contact.
4. Attend webinars and workshops that educate people about the TWI programs.
5. Attend the annual TWI Summit Conference (www. twisummit.com).

Select a Pilot Project—To Show the Need for Standard Work

Without standards, there can be no improvement.

Taiichi Ohno

Having standard work but not being able to train people on how to perform jobs in conformance with the standard is the same as not having standard work at all. There is no better place to begin introducing TWI than where there is a compelling need for standard work. If the initial project is selected well, the TWI effort will show immediate results. By stabilizing an unstable process, variation will be removed and problems will go away or rise up to the surface, where they can be spotted and quickly taken care of. With quick results, the TWI implementation will begin to gain momentum as people throughout the organization see what good skills training can do both for the operators and for the company.

Select a small project to showcase JI and then to grow the implementation. This does not necessarily mean that the amount of training will be small, but the target of the training itself should be limited in scope. For example, you may want to select one line, or one specific operation or set of operations, or one area that is causing problems. The target should be well known as being a trouble spot so that, when corrected, people will notice the difference. Even then, take good data before and after the training so that the results can be broadcast to show how TWI contributed to the improvement. Follow the TWI activities closely to ensure they are being carried out faithfully and energetically by those who have been properly trained in the ten-hour classes.

If you picked the correct area to run the pilot and if you shepherded the process meticulously to ensure that the TWI methods were properly installed as described in the next section, then before long everyone will see positive change coming out of the effort. Success will create "pull" from the entire organization for JI training as executives as well as supervisors begin asking for the training to be given in other areas outside of the pilot site. As Patrick always tells his new trainers, "You will know your efforts are succeeding when supervisors come up to you and ask to be put into a ten-hour training class." This will also be true for JR if it is given simultaneously as recommended for supervisors involved in the pilot. As we discussed in Chapter 5, when JR is given together along with JI, there is a much better chance of having good results.

The following success stories presented in this book began with good pilot projects that showcased TWI as discussed above:

■ Nixon Gear (Chapter 2): Introduced JI after a kaizen event to standardize work in the precision compressor gear cell, which was the most productive cell in the plant at the time. The improvement team created Job Instruction breakdowns and then retrained operators on the new standard. The cell realized a 17% increase in throughput with less physical stress on the cell operator. By improving what was viewed as "the best process" employees' eyes were opened as to what could be done to improve other operations within the plant.

■ Albany International Monofilament Plant (Chapter 6): Standardized work for the newly created set-up operator position as the key to improving through-put. Created a set of JI breakdowns on critical tasks to identify and then teach best practices, which resulted in a 70% reduction in human error.

■ Virginia Mason Medical Center (Chapter 9): Standardized the method for hand hygiene, a classic example of targeting a very visible problem to showcase the value of JI. The new method improved reliability in hand hygiene from 83.5% to above 98% and created "pull" for training in the new hand hygiene method from the entire organization.

■ Currier Plastics (Chapter 11): Work standards developed, beginning with the Tottle inspection project, after front-line supervisors in the Blow Molding Department were trained in JI and then trained their people using TWI. As a result, the cost of quality was reduced from 4% to 2% between 2007 and 2008, increasing overall efficiency by 3%. TWI was then introduced to Injection Molding where JI is being put to use to reduce quality issues and to improve productivity in that department.

Initial Delivery of TWI Training

By maintaining a limited scope to pilot the initial TWI effort, you will be able to avoid the big mistake some companies make of focusing on instructing all the jobs in a department or plant at once. As tempting as this may sound, such an effort is doomed to failure because the training activity is so dispersed. While some good results may happen, the training cannot be monitored and controlled and will soon get watered down or lose momentum. Start small to *build momentum* for sustained change over the long run.

No two companies are alike, and no two plants even within the same organization have the same culture. Therefore, there is no set plan to begin TWI training and it is best to start out slow to speed up the process as you roll TWI out and learn by doing. But some considerations for the initial training classes can be as follows:

■ The first class of trainers should be a select group of people who will be responsible for implementing TWI in the pilot area. The mix of people should include the trainer candidates (those expected to become ten-hour class trainers), supervisors, team leaders, informal floor leaders, operators, support staff, and union representatives.

■ The JI and JR programs should be delivered in roughly the same time period to a class of ten participants. (See Chapter 3 for this timeline and why it is recommended that a company should start with JI and JR.)

■ Training times should be flexible to allow participation of people who work on different shifts.

■ If they are not actual participants in the class, key people, such as management and staff, may and should observe the training to understand the content. Only then will they be able to support the people being trained when they move on to implementing the TWI skills. Such observers are not to interrupt or take part in the class activities.

■ TWI trainers should provide guidance for participants to select a practice demonstration for each class.

■ To a limited degree, TWI trainers also give additional instruction and guidance with making JI breakdowns for their practice demonstrations in the workplace. But they should never tell them how to do it. That would take away from the learning experience.

■ While all of the practice demonstrations take place in a classroom setting where people can hear with few distractions, we have found it very effective, if time permits, to do one *last* application practice on the plant floor using a real operator who has not taken the JI class to act as a learner. This allows the class to get a taste of what JI is like in actual practice, now that they have learned how to do it in the classroom.

The people who take these initial classes should then begin applying their new TWI skills to the jobs in the designated pilot area. The TWI working group leading the project should be completely engaged in this process so that questions are answered and stumbling blocks overcome. For example, as we saw in several of the case studies, there will inevitably be much difficulty making the initial JI breakdowns for training, as this skill takes a lot of practice and perseverance to master. In the W. L. Gore and Virginia Mason Medical Center examples, the teams of trainers spent a lot of time and effort up front developing the breakdowns and making sure they were "just right" with trial deliveries before taking them out to the floor to train. This investment will pay great dividends when it comes to successfully training the jobs, and this will in turn lead to good results.

Having made sure that the TWI methods were actively applied to the pilot area, positive results should follow as the targeted jobs begin to stabilize. Because we have picked a pilot area that is well known for troubles, everyone in the organization will see the improvements made through the TWI intervention or, if they don't see them, you should broadcast these results using the before and after data you meticulously collected during the pilot period. It won't be long, if it hasn't happened already, before other areas begin asking for the TWI training to be given to them so that they, too, can benefit from the method and get the same kind of good results.

Based on this feedback from the organization and your strategy for wider implementation, you can begin to plan the next areas to apply TWI. Keep in mind, as you move forward from the pilot, that it will still be wise to "go slow to go fast." Resist the temptation to apply the method everywhere but, taking it one step at a time, continue to build on the pull you have created from the overall organization. The moral of Aesop's tale of "The Tortoise and the Hare" applies especially well here— slow but steady always wins. (As Patrick's operations professor told his class many years ago, "Aesop would have made a hell of a production manager!")

Develop Certified In-House Trainers and Spread the Training—The TWI Institute Certified Trainer Programs

Once the pilot has had time to show results and nears completion, it is time to think about the next step, which is to have more people outside of the initial pilot group take the ten-hour training and begin practicing the TWI skills. You could have certified TWI trainers come back and deliver more sessions, but the best way of proceeding is to develop in-house trainers who take a trainer development program, sometimes called a train-the-trainer class, in which they learn to deliver the ten-hour sessions following the TWI trainer delivery manual for each module. These programs were designed to spread the training and engage what is called the "multiplier effect."

"When a program is being operated nationally, quality control is necessary so that there can be a guaranteed quality standard, nationwide."[*] The founders of TWI recognized that, in order to meet their wartime production goals, they had to create a program that was universally applicable and proven to provide success each time it was delivered. In Japan, after the war and until this day, the Japan Industrial Training Association (JITA) has continued to promote TWI just as they were taught by their American predecessors. Based on Patrick's experience developing TWI trainers at SANYO, where they followed the JITA pattern and leadership, the TWI Institute has modeled its TWI trainer development programs following the very same format that was used by the TWI founders and transferred to the Japanese. In fact, Patrick's

---

[*] TWI Report, 1945, page 178.

TWI mentor, Kazuhiko Shibuya, learned to deliver these programs as a young training staff member at SANYO where they ran TWI trainer sessions led by Kenji Ogawa, one of the original Master Trainers taught directly by the American TWI trainers in the 1950s.

Having trained more than 400 trainers since 2002, the TWI Institute has demonstrated that the success of this program is predicated upon strict adherence to proven TWI training methods and use of proven TWI training materials. In order to do this, trainer candidates first take a TWI ten-hour class and become proficient in the TWI skill they have learned. Then they spend a full week with a Master Trainer practicing the delivery of the course following the instructor manual which outlines every aspect of the ten-hour class including all material presented, demonstrations, discussions, board work, and so on. The manual literally provides a "script" for the trainer to follow so that the content is presented consistently each time it is delivered. For the most part, this is the same "script" that was formulated by the TWI founders when they created the program in the 1940s.

The value of being a "Certified TWI Institute Trainer" lies in ensuring that the delivery of the program is consistent with the successful presentation of the material over six-plus decades. As in any job, just because the procedure is written down in a "manual," doesn't mean that the work can be done simply by reading that manual. There is a great deal of nuance and technique that can only be acquired through practice under the direction of a person who has deep experience in both teaching the class and developing other trainers to teach it. Under this methodology, even inexperienced people can become excellent TWI trainers by learning and mastering the delivery following the training manual.

There are many reasons why this is the best approach, but perhaps the most important one is that in-house trainers can act as the TWI experts for the company in the modules they are teaching. As we heard Mark Bechteler explain in Chapter 2, taking the trainer development courses helped him to get much deeper into the material and to understand the nuances of each method as he prepared to teach others how to use them. Once they begin training others and help them begin practicing the skills for themselves, the trainers' mastery of the method grows. As the saying goes, the person who learns the most in the class is the teacher. The TWI trainer who works inside the company and knows the many details of the work of the company is in an excellent position to guide the effort as it grows from section to section and from department to department. An outside trainer will not be as familiar with that work and will not be as available on a constant and consistent basis.

The next question that comes up is: "Who is best qualified to teach TWI?" The TWI trainer delivery manuals were designed and written so that even a person who had little or no experience teaching classes could learn how to put on the courses. They literally tell the trainer what to say and do every step of the way. But it is a big mistake to think that one can simply pick up the manual, study it, and then begin teaching classes successfully. We saw how the Japanese Labor Ministry, in its first attempt to get TWI started in Japan, failed to trigger the multiplier effect because it thought that a single person could go learn it and bring it back. Fortunately for everyone, the Japanese wound up starting over by hiring qualified instructors who were able to train trainers properly based on their vast experience developing and promoting TWI during the war. There is an incredible amount of wisdom and technique embedded in the TWI manuals, and it takes an experienced TWI professional to guide a new trainer to finding out how to deliver the courses correctly using this invaluable resource.

People with no training background, when put into a trainer development program directed by a qualified master trainer, can learn to teach the TWI classes effectively, getting the same great results literally millions of trainers have been getting for over six decades. In fact, our experience has been that the best trainers are those who have little or no training experience but a deep background in the industry and jobs they are working in. This is not a coincidence. On the one hand, consultants and professional trainers tend to want to put their own flavor or spin on the training they give by injecting their unique personalities and perspectives. This goes directly against the standard delivery approach of the TWI formula, and these people have a difficult time when they are told to follow the manual. On the other hand, because the bulk of the TWI ten-hour classes are spent on actual jobs and problems brought in from the work site, the trainer must be familiar with the work being done in the company to effectively lead the group in applying the TWI methods in this learn-by-doing approach. Understanding the jobs, then, is a more important qualification than experience as an instructor.

With in-house trainers on board, the company can put on TWI sessions at will and where needed, developing these essential skills throughout the workforce. These trainers will also be able to serve as coaches and consultants for the ongoing TWI implementation as it spreads throughout the organization.

The Rollout Plan

We started out this chapter showing how C. R. Dooley of the TWI Service instructed companies on how to get continuing results in a plant from Training Within Industry in June 1944. As he said so eloquently, "It is another splendid example of the combined efforts of many TWI people." While there is no cookie cutter approach to rolling out TWI, we feel that the best plan that provides the foundation for companies to build a long-term successful TWI program is just the Phase 1 and Phase 2 approach Dooley gave us, as outlined at the beginning of this chapter. It is only when companies deviate from this standard that they fail, as was experienced by Nixon Gear, when they originally did not follow up with the training, and by Albany International, where people did not learn how to properly break down jobs using the JI method.

In addition to these valuable guidelines, *The Toyota Way Fieldbook* also serves as a valuable guide to the sequence of events Toyota traveled to teach their people to use the TWI tools in the Toyota Production System. Making respect for people (Job Relations) and standard work (Job Instruction) part of a continuous improvement (Job Methods) program is definitely a big step in the right direction. More than just a philosophical approach, Toyota puts its ideas into actual practice, as can be seen in how it deploys its people to implement these ideals:

> Toyota has a system of group leaders and team leaders. The team leaders are hourly and are responsible for supporting about five to seven associates. They audit the work procedures of employees to detect deviations from standard work … and provide the necessary structure to fill in for absences. They are often involved in developing standard work for new models. They are a key to turning standard work from good looking wall hangings to true tools for continuous improvement. Interestingly, the team leader role is exactly what is missing in most companies.[*]

Table 10.1 sums up these implementations steps. By getting passionate people in place to lead a TWI implementation, by creating a good plan to start training, by being sure management's strong support of the TWI skills is firmly in place, by creating in-house trainers who will continue to spread the program and follow up directly with implementation, and by supporting training and TWI activities on the floor, you can be sure to have TWI fulfill its promise of creating a strong foundation for Lean.

How to Get Continuing Results from JI, JR, and JM[1]

In the course of implementing Lean manufacturing across the country, we have noticed that many companies have

---

[*] Jeffrey K. Liker and David Meier, *The Toyota Way Fieldbook*, McGraw-Hill, New York, 2006, p. 117.

Table 10.1   TWI Implementation Steps

Form TWI working group	Select strong leader
	Form cross-functional team
	Study TWI fundamentals
Create pilot project	Limited in scope
	Line/operation/area in need of standard work
	Obtain current metrics
Initial TWI training	Ten-hour classes of JI and JR
	A select group who will work in pilot area
	Include TWI trainer candidates
Carry out pilot project	Break down and train pilot area jobs
	Close guidance and support from TWI working group
	Obtain after-training metrics
Publicize pilot and plan next steps	Create pull from other parts of organization
	Plan for next areas to push out training
Develop in-house trainers	Key members learn to teach JI/JR ten-hour class
	Training for supervisors outside of pilot area
Roll out TWI to entire organization	Slow but steady—not everywhere at once
	Ensure top management commitment
	Provide adequate coverage
	Continued coaching for supervisors
	Reports to management
	Give credit for results
Introduce JM	When ready, introduce and promote JM
	Create system for handling improvement proposals

difficulties sustaining results from their Lean effort due to the lack of a good understanding around how to manage the daily work routine. Efforts to create leader standard work have gone a long way to address this problem, and this is a step in the right direction. But supervisors still need to understand their role as the point person in a smoothly run organization, and how this role determines what they should do every day.

Generally speaking, most supervisors regard the management of daily routine as an activity that carries them from one problem to the next. The day goes something like this:

> I arrive at work and try to get a feel for where things stand. I check the incoming queue of jobs to see what's still there from yesterday, what's new, if there any hot jobs, if there is anything that looks like it has problems, etc. Then I check on my people. Have they all come to work? Are they working or socializing? Are they where they should be? By this time I usually have gotten a phone call or an e-mail, and I've been asked questions on status, or I've been told something I have to do, or an employee has pointed out there is something wrong with the job he or she has to do. It doesn't take much time to have several problems appear for me to start working on.
>
> It pretty much continues that way for the whole day, and then I go home. Maybe I was able to do a little planning and thinking about what has to happen tomorrow or next week, or I might have started working on my weekly production report—but that's if I'm lucky. Most of the time I'm just on the go, dealing with things as they come up while things that I consider important don't get done because I just don't have the time.

It doesn't have to be this way. Lean manufacturing teaches us that there are specific elements of a world-class management system. But it too often takes companies many years of experiencing the difficulties of implementing Lean to see that they, in fact, don't fully understand how these components work together to dictate a culture of *daily activities* that guide front-line supervisors toward achieving their goals. Let's take a look at the background and then develop how TWI fits into this system to provide the skills needed to properly and effectively carry out these daily activities.

A World-Class Management System

We have had a few decades now of studying and learning what it takes to make a management system that can compete with the best competitors in the world. While there is no shortage of ideas and opinions in today's Lean manufacturing world, we can boil the key elements of that system down to a fairly simple model. A world-class management system has three interrelated components (see Figure 10.1):

1. Policy deployment: Aligns the entire organization with long-term and annual objectives.
2. Daily management: Creates the structure for putting standards in place and continuously improving.
3. Cross-functional management: Manages improvement of the entire system for innovation and breakthrough.

Policy deployment refers to the various methods used to be sure that everyone in the organization is working effectively toward the same ends. While some statements of policy are administrative and meant to guide routine decision making, policy deployment includes the creation of policies used to manage and guide improvement of the company situation. In other words, it is not just the actual written policies, but the process by which those

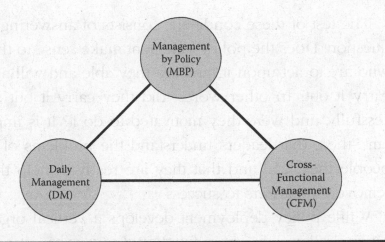

Figure 10.1   A world-class management system.

policies are created and put into place. In this way, the organization is always a growing and evolving entity that is not tied down to a stagnant and fixed set of rules.

Deployment refers to the manner in which the policies are communicated. To be effective, policies must be communicated without ambiguity. They must be workable and understandable by those who are to carry them out. When members of the organization are unsure or tentative about their roles in making these policies work, then the chances of those policies succeeding are put into question. To ensure that policies have been effectively deployed, then, the communication of these policies must be *tested* for comprehension.

To ensure comprehension, two conditions must be met:

1. You do not know how someone else has understood what you have told him or her until you see how he or she interprets your statements to someone else.
2. The spoken word is inadequate for policy deployment. Policies must be written.

The test of these conditions consists of answering the question: Does the policy statement make sense to those who are to act upon it, and are they able and willing to carry it out? In other words, did they carry it out successfully, and were they motivated to do it? It is important, then, that leaders understand the problems of the people they lead and that they are ready to help them remove the barriers to success.

While policy deployment develops a vertical organizational alignment, *cross-functional management* refers to a horizontal alignment that crosses the functional boundaries of an organization and prevents suboptimization. For example, what looks like a good improvement for one department may actually not be good for the company as a whole. In addition, cross-functional management fosters the necessary collaboration to overcome functional barriers when there is a need to innovate in support of long-term and annual improvement goals.

Daily management refers to the management of the work routines—the jobs—performed in the company in such a way as to achieve and sustain basic stability through standardization. This also includes modifying the standards to address abnormalities when they arise and to improve on those standards in order to achieve goals set by the organization. If the work of the company is not performed in a consistent way, always maintaining quality and meeting other standards of completion time, safety, cost, etc., then there is no guarantee that the company will meet its bigger goals of profit and growth. The role of the supervisor, then, is to manage this work on a daily, continuous basis.

The PDCA Cycle

Embedded throughout this management system is a method to obtain results. In other words, the interrelated parts of the management system by themselves will not ensure that goals are achieved. They need to be actualized by specific steps to ensure they function as designed. The method of management is the plan-do-check-action (PDCA) cycle, and it bears saying that all four steps of PDCA are equally important. Performing just one or some of the steps renders the method quite useless. It is essential to cycle through PDCA to ensure continuity in the management process by repeating the cycle over and over again.

- *Plan* is the basis for execution and checking. It consists of setting goals and targets, defining measurements that will quantify progress, setting the time limit for achieving the goals and targets, and creating a definition of how to arrive there—the means.
- *Do* is to carry out what has been planned. Training people in what they must know, understand, and be able to do is always part of the *do* step. Even the best plan will fail if people lack the necessary skills and competencies to carry it out.
- *Check* is to study the outcomes of what has been done and to compare these results with the stated goals.
- *Act* is the response to what has been learned in the *check* process. If the results obtained were those that were planned, then the *act* is to standardize. If the planned results were not obtained, then the *act*

is to refine the plan and repeat the cycle until the outcomes turn out as intended.

The three elements of the management system—policy deployment, cross-functional management, and daily management—integrate with the management method of PDCA and make up the whole process to realize the concept of continuous improvement. In order to achieve the standard goals (production output, quality, cost, safety, and morale), management must first establish work standards, actively train people in those standards, monitor and ensure that the standards are consistently applied, check on the results produced, and modify the standards when necessary to continue to meet standard goals. This is sometimes called standardize, do, check, act and is truly the crux of effective management. Then, once work standards are firmly established, we continuously improve on those standards, using the PDCA cycle, through incremental improvements that eliminate waste and non-value-added elements and processes. Finally, through cross-functional collaboration, we innovate and "raise the bar" in terms of competitive performance.

Although the daily management system is linked to policy deployment through a focus on both the routine and improvement, there is still a gap. Theoretically, everything in these Lean models makes sense, and companies implementing Lean manufacturing find them easy to understand. But even though they understand them and want to apply the model, they never seem to take root in the "firefighting" culture we find in almost every industry and every company. So they just remain theoretical concepts that everyone agrees with but can't do.

The PDCA cycle begins to truly turn when we learn and implement TWI. The gap, of course, is that without important competencies fully in place, people just don't have the ability to follow through effectively on their roles in the management system, and therefore cannot sustain the changes they make. Implementing TWI, just as we have been describing throughout this book, and getting experience using it, makes it clear that the daily management component is supported and made effective when supervisors and managers have developed the skills to:

1. Instruct effectively in what the operational standards are and how to follow them (Job Instruction)
2. Know how to study and analyze operations to either:
   a. Establish standards where there are none or improve the ones in place when there are abnormalities (Job Methods)
   b. Actively find better ways to meet standard goals with less waste (Job Methods)
3. Lead people effectively so they can contribute fully to changes being made and remain motivated to follow and sustain new standards of operation (Job Relations)

Making TWI Part of the Daily Management System

Creating a skills-based culture means abandoning the reactive practices of the firefighting mentality that robs our organizations of the real attention and effort we should be directing toward making the management system work. If we are to achieve the goals that will propel us into the future, we must create the discipline to follow time-tested practices, like PDCA, that create stability and continuous improvement. To do that, we need basic skills that we can deploy on a *daily basis* to maintain those good practices. By leaning on these skills as what supervisors do every day, we can better ensure that the fundamentals of stability through standard work, motivation of the workforce, and continuous improvement become the culture of our working environment and how our people behave on a regular basis.

The three-legged stool analogy that represents the TWI program and its three modules—Job Instruction, Job Methods, and Job Relations—can be used again here (see Figure 10.2). The three legs of TWI firmly support

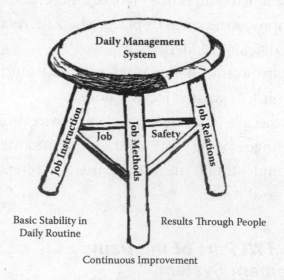

Figure 10.2 Daily management and the three-legged stool.

and hold up the daily management system. If we also consider a fourth element, Job Safety, which was developed in Japan in the 1950s, as an integral part of all three TWI skills, we can see the complete model of the skills set supervisors need to use every day to create and maintain the culture of competence we have been describing in this book. As the many case studies we have looked at show us, when companies train their front-line management in these leadership skills, they begin to gain control of their work, instead of the other way around, where they are chased by their work on a daily basis. This control is what the management system seeks to leverage in order to attain the organization's goals.

Endnotes

1. *How to Get Continuing Results from JI, JR, and JM* and *A World Class Management System* was contributed by Richard Abercrombie, president of Lean Promotion. Richard has over twenty years of experience in manufacturing, most of which was obtained while employed by the Boeing Company in Washington State. There, Richard became involved in Boeing's Lean manufacturing initiative as a member of the Boeing Supplier Support Center, which assisted key Boeing suppliers, both domestic and international, with beginning a continuous improvement program based on the Toyota Production System (TPS). To fulfill this role, Richard received extensive training in TPS from the Shingijutsu Company both in Japan and at Boeing facilities in the United States.

   Richard has significant experience delivering these TWI programs around the world as an integral part of his TPS consulting for clients in the manufacturing, aerospace, steel, food processing, banking, construction, and healthcare industries. Richard was the first master trainer trained by Patrick Graupp to train new trainers in each of the forty-hour train-the-trainer program modules of JI, JR, JM, and JS, as Patrick was trained to do in Japan. His strong background in TPS consulting is put to good use when Richard collaborates with the institute to deliver follow-up coaching for companies to quickly learn how to implement TWI as an integral part of their Lean strategy.

Chapter 11

A Preemptive Turnaround at Currier Plastics, Inc.

Currier Plastics, Inc. was established in 1982 by Raymond J. Currier as a custom injection molder of thermoplastic resins servicing a broad OEM customer base. Ray's vision was to offer superior engineering and comprehensive customer service as the foundation to long-term customer relationships. He began with two injection molding machines, four employees, and a handful of local customers in 4,000 square feet of converted warehouse space located in Auburn, New York.

Under Ray's leadership, Currier Plastics, Inc. (Currier) grew to become recognized as a preferred supplier of precision injection molded components for companies in the automotive, medical, telecommunications, and consumer products industries. Ray retired in 1987, confident that his sons Jim and John would continue to grow the company in accordance with their shared vision. Since then, the original facility has been expanded several times adding buildings, equipment, and personnel. Extrusion blow molding was added to the company's core competency in 1996, allowing it to leverage its existing knowledge of part design and plastics processing into a more targeted market segment that purchases both injection molded and blow molded products. In 2002, Currier added a product design function to establish a market focus on packaging products, a unique combination of molding processes that very few molders in the United States offer.

In the late 1990s and moving into the new century, Currier experienced double-digit sales growth for seven straight years, but such dynamic growth came with a price. The company experienced a high turnover rate for technicians, operators, and temporary employees that contributed to a high material scrap rate, high cost of quality, excessive machine downtime, and low efficiencies that could be traced in part to obsolete work instructions for most jobs. The company was not keeping up with the growth of its business.

Three of the four current owners that work full time for the company experienced these day-to-day challenges involved with growing the business, and in 2005 decided that Currier needed to do things differently to succeed as a larger company. Potential vendors were invited to come and talk with them about strategic planning. One of these vendors, Thomas Walsh, PhD, with the Grenell Consulting Group, had this to say about his initial meeting:

> My first meeting was with the Currier Plastics owners, board of directors, and managers—group think was prevalent.... While this is not uncommon in small companies and professional services firms, it

> usually presents problems of organizational clarity
> and professional boundaries and makes it hard for a
> CEO to really lead.*

Although Currier had done strategic planning in the past, that plan was left sitting on a shelf and was never used to drive the business. It had always been a company that rolled up its sleeves rather than sharpened its pencils when it came to getting the job done. Rather than jumping in and creating a new plan, then, CEO John Currier decided to first do a leadership audit, as recommended by Dr. Walsh, to determine the strengths and weaknesses of the management team in order for the company to "get the right people on the bus, and in the right seats, before the bus left the terminal." John did not want to embark on yet another dead-end journey.

Going through this process led John to realize that he needed to hire a qualified person for business development so that he could spend more time focusing on being the CEO. The entire leadership team also went through a leadership inventory that resulted in a number of changes. Some people wound up leaving the company, while the roles of others were changed to get fresh ideas and perspectives. In early 2006, he positioned Max Leone to become Vice President of Business Development, and along with Alan Gross, the new Vice President of Operations, the company's refurbished executive team began to pursue profitable growth and a culture of continuous improvement.

---

* Thomas Walsh, "It Starts With The Team: A Preemptive Turnaround at Currier Plastics," Grenell Consulting Group, *CNY Business Journal*, November 13, 2009.

They immediately developed a strategic plan for Currier that focused on a target market of companies that required the integration of design expertise: injection molding *and* blow molding (see Figures 11.1 and 11.2). Sixty percent of Currier's current customers purchase both injection and blow molded products, allowing the company to single source their requirements and insure design integration for manufacturability. Investment in technology in both design and manufacturing is a key aspect to the continuation of this plan. In addition to customer-oriented goals, the strategic plan sought to improve the quality of work life for the company's employees, with the goal of making Currier Plastics the best place to work in the business.

This chapter is about how the company achieved the goals of its strategic plan using TWI as a critical part of that

- 14 Machines

- Container size 0.5 oz to 1 gallon

- Materials
  HDPE
  LDPE
  PP
  PVC
  PETG
  PET

- Integrated automation & leak testing on equipment

- SPC inspection station

- Climate controlled

Figure 11.1 Blow molding operation. (Reprinted with permission of Currier Plastics, Inc., Auburn, NY.)

- 24 Machines
- 20 to 500 US tons
- Part size 0.5 g to 30 oz
- Engineered thermoplastics (over 50 active resins)
- Integrated automation on equipment
- 3 Quality test stations
- Climate controlled

Figure 11.2 Injection molding operations. (Reprinted with permission of Currier Plastics, Inc., Auburn, NY.)

developmental process. Today, Currier Plastics' unique value proposition of "providing custom packaging solutions that leverage core competencies in design, injection and blow molding" is a competitive niche in the industry. The company's key market segments in the packaging industry include personal/healthcare, beauty and cosmetics, household chemicals, and amenities. Currier is best known for wet-wipe canisters/lids, and amenity products for high-end hotels and resorts. Other key markets include consumer, medical, and electronics products. The company now employs over one hundred people in a 65,000 sq. ft. state-of-the-art molding facility, in its original location in Auburn, New York, where all employees take pride in the company having been awarded the Association for Manufacturing Excellence's 2009 Mid-Atlantic Regional Manufacturing Excellence Award in October 2009.

A Preemptive Turnaround— 2006–2009[*]

Alan Gross had graduated *summa cum laude* with a BS in industrial engineering from the University of Buffalo, and continued his education taking graduate courses throughout his seventeen-year career with Kodak in Rochester, New York. He also taught in the highly regarded Masters of Leadership program at the Rochester Institute of Technology. Alan left Kodak and went on to lead the transformation of a midsized plastic injection molding company into a multi-award-winning organization with unprecedented profitability in the plastics industry. The company was soon purchased by a larger organization that deemphasized plastic molding. Not wanting to uproot his family, Alan interviewed with Currier Plastics for the new position of Vice President of Operations, where he found a new home working with owners who were committed to staying in business and to providing good jobs for the community in which they lived.

With the executive level players now in place, the next step was to create a strategic framework beginning with the mission, vision, and core values of the company in order to establish company-wide alignment on achieving key objectives. These elements were established as follows:

Mission: Currier Plastics strives to be the best in the world at providing integrated plastics solutions from

---

[*] The bulk of the information contained here is taken from the Currier Plastics Application for The Association for Manufacturing Excellence 2009 Mid-Atlantic Regional Award submitted May 31, 2009.

design to final product, creating unequaled value
for our customers and limitless opportunity to our
coworkers.

Value Proposition: Integration of the core competencies of design, injection molding, and blow
molding.

Core Values: Currier Plastics strives to create a culture
of integrity and mutual respect for our coworkers,
our customers, and our community. We exist to provide value and outstanding service to our customers.
We pursue ever-higher goals by embracing change
and fostering an environment that values passion,
creativity, and individual dreams.

Developing a Strategic Framework for Improvement

The Currier Strategic Framework identified three Key
Results Areas that would establish the framework for their
efforts moving forward: Talent Management, Velocity ×
Value = $V^{2©}$ (increasing rates of improvement in quality, cycle time, and cost), and Innovative Growth (see
Figure 11.3). These three areas represented customer-
driven factors their strategic planning process told them
would be critical to success. The plan also identified key
measures that would track results in each of these areas,
along with the strategies for improvement that identified
major initiatives toward achieving these results. Finally,
specific Improvement Projects were outlined for each
initiative and their objectives further refined in one-
page plans that were linked from the Corporate to the
Department levels (see Figure 11.4).

Currier Plastics, Inc.

Customer-Driven Strategic Framework

Key Results Areas

What are our critical customer-driven success factors?

Talent Management *Leader: CFO*	*Velocity × Value = V²* *Leader: VP Operations*	*Innovative Growth* *Leader: VP Business Development*
We constantly strive to strengthen the individual and team performance in a way that will allow Currier Plastics to be recognized the best place to work in the world.	Exceed world-class standards, leading to continuing rates of improvement in quality, cycle time, and costs so that we can offer increasing value to our customers and are recognized as the benchmark in the industry by all definitive measures of success.	Leverage innovative Currier Plastics core competencies to grow profitably in order to ensure increasing development for our customers, employees, suppliers, owners, and community.

Figure 11.3   Strategic framework 1. (Reprinted with permission of Currier Plastics, Inc., Auburn, NY.)

Key Results Measures

What are our vital signs that indicate success?

1.1 Quality of work life	*Quality*	3.1 Capacity utilization
	2.1 Cost of quality	
1.2 Employee turnover	*Cycle Time*	3.2 Percent new business
	2.2 Total inventory	
1.3 Accident incident rate	*Delivery Performance*	3.3 Capability alignment
	2.3 On-time delivery	
	Financial	
	2.4 Return on production	
	2.5 Overall efficiency	

Figure 11.3 (continued) Strategic framework 1.

Key Results Strategies

What are our major initiatives to ensure success?

1.1 Create and nurture a culture of coaching, mentoring, and training to enhance individual value one employee at a time	2.1 Apply Lean thinking across the company to drive to world-class time competitiveness	3.1 Improve the product quality planning process
1.2 Obtain, develop, and retain the best people in all positions, focusing on both skill and talent	2.2 Proactive supply chain management (SCM)	3.2 Design, develop, and manage business development via a comprehensive business development plan consistent with corporate goals
1.3 Create a safe work environment and culture	2.3 Improve cash flow and financial rate of return	3.3 Investigate tool building and repair partnerships
	2.4 Optimize molding processes through use of scientific molding principles and statistical process controls	3.4 Continually strengthen our level of partnership with key customers and new prospects

Figure 11.3 (continued)   Strategic framework 1.

Key Results Strategies Improvement Projects

How do we reach our goals?

1.1.1	Recruit and hire technical talent that can facilitate corporate goals	2.1.1	Targeted kaizen and Six Sigma events	3.1.1	Create and nurture a culture of coaching, mentoring, and training to enhance individual value one employee at a time
		2.1.2	Integrate skills continuum and Training Within Industry methodology		
1.1.2	Offer and promote individual opportunities through a skills-based approach providing specific career paths	2.1.3	Total productive maintenance	3.1.2	Restructure business development and up-front engineering
		2.1.4	Implement Lean SCM that incorporates improved flow and minimizes nonvalue activity from suppliers to the customer	3.2.1	Implement business development plan designed to achieve growth objectives
1.1.3	Develop value-added plans in response to QWL information gathered	2.2.1	Make better use of Vista to provide improved costing information, forecasting, scheduling, and inventory accuracy	3.3.1	Develop a worldwide network of tool design and tool builders focused on cost and time competiveness
1.2.1	Increase bench strength through development of out-of-the-box recruiting, training/personnel development	2.2.2	Molding machine modernization	3.4.1	Establish, maintain, and disseminate a formal customer intelligence database and establish quarterly reviews of customer strategic fit, contribution, and maintenance
		2.3.1	Tool room development		
		2.3.2	Complete cell within a cell concept development		
1.2.2	Establish turnover metric that can be competitively benchmarked	2.3.3	Apply scientific molding (decoupled) and Six Sigma, automatic process control to existing production, and quality planning for new products		
		2.1.1	Expand visual management system		

Figure 11.4   Strategic framework 2. (Reprinted with permission of Currier Plastics, Inc., Auburn, NY.)

Talent Management Creates a Flexible and Engaged Workforce

Talent management, the first Key Results Area, is an extension of the company's "right people in the right places" objective. What was achieved first at the executive level was then applied to all subsequent levels of the company. Talent assessments were applied using an A/B/C analysis similar to the practice GE applied during the Jack Welsh era. Then, a nationwide search was conducted using a variety of recruiting avenues. Team interviews were done in a "gates and phases" way, aided by a proprietary personality profile match to the position being considered. Those determined to be at an A level, or with the potential to reach that level, were actively recruited, mentored, and coached.

As for training and developing people already in the company, a career development specialist was put in place, and numerous grants were obtained to develop the capabilities of everyone in the company. Different training approaches were used at different levels of the company. For example, skilled positions followed a promote/develop/coach from within philosophy.

Using a temp-to-hire strategy (new hires are hired as temps for evaluation before being hired as full-time employees) developed during a Kaizen Event, Currier certified preferred temporary employment agencies and equipped them to effectively orient and train their people for work in Currier's particular work environment. An experienced plastics engineer was hired to develop "striped unicorns," the Currier term for the very rare process technicians skilled in Continuous Extrusion Shuttle

Blow Molding—their specialty. An improved performance appraisal process was developed, and training was provided for all supervisors and management. This new appraisal form was derived directly from updated and comprehensive job descriptions for all jobs in the company. Merit pay and bonuses were paid based upon a mix of individual contributions, departmental results, and company-wide performance.

The company began applying TWI in manufacturing and formulated it to be a never-ending process to develop and sustain fully interchangeable, self-managing work teams. Their approach is challenging since every employee on the shop floor is now a target of the TWI training and follow-up. How they fully integrate TWI with their lean strategy will be presented in more detail later in this chapter.

As can be seen from the results in the following productivity summary (Figure 11.5), Currier was able to better

Metric	Units	2006	2009	Change
Sales	Normalized	100	111	11.0%
Annual profit	Normalized	100	729	629.0%
Cost of quality	Percent of sales	6.3%	1.7%	−73.0%
Efficiency	Percent possible	76.5%	90.2%	17.9%
On-time delivery	Acknowledged	88.8%	99.5%	12.0%
Total inventory	Days of supply	15.5%	11.5%	−25.8%
6S audits	Percent possible	33%	81%	145.5%
Quality of worklife	Survey results	71%	84%	18.3%

Figure 11.5 Key results summary. (Reprinted with permission of Currier Plastics, Inc., Auburn, NY.)

establish a continuous improvement culture by developing a fully flexible workforce, each employee "spontaneously driving improvement in the company without boundaries." In addition, everyone at Currier now feels more fully engaged and valued within the organization, though the challenge now is to maintain the momentum of this progress.

Applying Velocity × Value = V²© across the Company to Exceed World-Class Standards

V^2 means doing everything in a *quality way, faster than others* to create a competitive advantage where value is defined from the customer's point of view. For Currier, value takes the form of unique and innovative designs for packages that few others can effectively copy. In fact, a large part of their value-added contribution is from engineering design for manufacturability and assembly. In addition, V^2 is their term for operational excellence, which is achieved by eliminating waste in accordance to the process flow diagram shown in Figure 11.6. This process is now a daily passion at Currier Plastics, but that did not occur overnight.

Innovative Growth to Secure the Future

The third and final Key Results Area is Innovative Growth. The key factors include:

- Well defined criteria used to evaluate the customer on a job-by-job basis.
- A New Product Development team that has a visual control center (war room) where new product

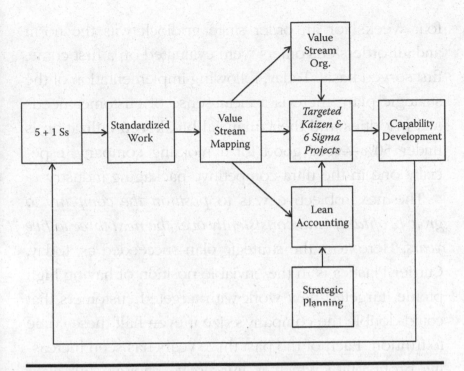

Figure 11.6   Lean enterprise V² process map.

milestones, design projects, and targeted new prospects are reviewed twice weekly.

■ A comprehensive database used to assess strengths and weakness, compared to the competition and from the customer's perspective.

■ A system to identify the latest technology and develop a capital plan to purchase, modernize, or upgrade existing technology.

Establishing Key Objectives

The first overarching objective was to *reestablish profitability*. In the past, machine utilization was close to 80%, so high it was unprecedented in the industry, yet the company struggled to break even. Lead time was over four weeks for an order since gridlock was the norm and all orders/customers were evaluated on a first come, first served basis. Today, following implementation of the strategic plan, there is a clear sense of customer needs and priorities and their financial breakeven utilization is under 50%—very good for a molding company, especially one in the ultra-competitive packaging industry.

The next objective was to *position the company to grow profitably and consistently over the next three to five years*. Here, too, the strategic plan succeeded as, today, Currier Plastics is in the enviable position of having high profile, targeted new work with targeted customers that could double the company's size if even half these come to fruition. Each of the past three years has seen increasing profitability which is impressive enough, but even more so considering the difficult economic conditions the entire country has faced in this time period.

Their final objective was to *become the employer of choice for all current and future prospective employees*. Quality of Worklife (QWL) surveys are given to all employees annually as the primary way of evaluating the success of the company on this objective and these have tracked the progress of the strategic plan. According to CEO John Currier, "We've seen a 10% increase in scores on that survey with favorable trends based on quarterly evaluations. Anecdotal evidence is equally strong—a full 10% of the current employment base includes those that left for greener pastures yet returned not long afterwards. These have turned out to be some of our most loyal and involved employees."

Implementing the Currier Lean Strategy

Alan learned from over twenty-five years of experience implementing change before joining Currier that the company would need a supportive outside partner to accelerate achievement of its objectives. With the Currier Strategic Plan in hand, he arranged for a meeting with Cindy Oehmigen, Senior Process Improvement Specialist for the Central New York Technology Development Organization (CNYTDO). Working with Cindy as their CNYTDO Project Manager, a sixteen-month Lean Implementation Plan was agreed upon that began with a four-hour overview of Lean Concepts for all Currier personnel. A core group was then organized and divided into five process improvement teams to focus on inventory reduction and process improvements across all shifts. Team members were provided additional training as needed in Value Stream Mapping, 5S/Visual Controls, Set-up Reduction, Total Productive Maintenance, and Standard Work.

Alan appeared to have everything in place to implement the lean strategy except for someone to drive change in the workplace. The need for additional process engineering support and for someone to take on the role of technical trainer led Currier to rediscover Dustin Dreese, who had worked for them as a process technician for a year after earning his BS in plastics and polymer engineering from the Pennsylvania College of Technology. Dustin left Currier to take a supervisor position with a blow molding company in Pennsylvania where he was introduced to Lean Manufacturing concepts. Returning to Currier, that experience enabled him to immediately take the lead role in implementing 6S and he soon became the Health and 6S Manager.

Six Ss (5S plus Safety)

Six S was the first step Currier applied toward eliminating clutter, by reorganizing and redesigning all its workstations. Daily responsibilities were defined at multiple levels for all to maintain the newly defined standards. Weekly local audits and layered monthly audits were used to verify compliance. Since the program's inception at Currier, the official 6S audit scores have increased 2.5 times—from 30% compliance to 75% across the whole facility—encompassing all areas: offices, blow molding, injection molding, and the warehouse. Alan proudly let it be known that "outsiders regularly compliment us on our plant and especially the organization and housekeeping. Hearing 'the best we've seen' has been a welcome and commonplace comment; it never gets old."[*]

Standardized Work and TWI

Cindy, who is one of three certified TWI Institute trainers at CNYTDO, suggested creating another team to provide supervisors and team leaders with the fundamental skills of Job Relations (JR), Job Instruction (JI), and Job

[*] Currier Plastics had learned so much from their plant visits to ESCO, Nixon Gear, and Albany International that they opened their doors to visitors from other companies looking to benchmark them.

Methods (JM) to sustain the gains being generated by the process improvement teams. Although Alan was not familiar with how TWI might fit into their Lean/Six Sigma strategy, he decided to evaluate TWI by having CNYTDO train a team of ten of his people on site.

The initial group to take the JI training class included key leaders and those expected to apply the techniques they learned to actually train people on the plant floor. Everyone involved knew the techniques were valuable and that this was a skill that would take a lot of practice to perfect. It was a complementary extension to the method Alan had applied during his time at Kodak, where TWI was used during World War II. Mentors were provided to help learn and apply the method. At the same time, the JR method was introduced to provide immediate value for supervisors who were inexperienced and new to their roles. Later JR became a keystone to the supervisory training at Currier.

A steering committee was formed to monitor the Lean/TWI progress—to confirm the direction it was headed in and to eliminate barriers to change. Eventually, this function was turned over to the natural steering team known as the Key Result Measure Team team (including all manager and department heads in the company). Benchmarking visits were instrumental in establishing a vision for what was possible and what barriers were to be avoided. The path to this implementation was made far less bumpy by learning from three nearby companies, which had experienced implementing TWI themselves, they visited and by customizing what was learned to the Currier vision already in place.

Currier's first Kaizen Blitz was focused on the area of most exposure and significant need—the blow molding of a new, high-volume "tottle" (a combination of bottle and tube used in the amenities market). This two-day effort resulted in two paths: a *technical path* focused on equipment effectiveness and a *human path* focused on successful processing, inspecting, and packaging of molded product. Also key to Currier's success was establishing a key internal resource as the lead for Lean applications. This role expanded into becoming roughly half of Dustin Dreese's job duties because it fit in naturally with his Process Engineering experience and his challenge to develop the next generation of technical resources in Blow Molding. Dustin was helped by experienced coaching from CNYTDO which was facilitated by grants obtained for this purpose. Later, these same grants, as well as others obtained, provided additional teachers, mentors, and consultants to the lean cause with a focus on TWI application.

Good progress was made with 6S as floor layouts became standardized and improvements freed up time for Operators and Lead Operators. The new Tottle process, however, provided many opportunities for improvement. Kaizen Blitzes identified readily solvable issues within cells but there was still a plant-wide lack of standard work. Many work instructions on the production floor were outdated, not being followed, or nonexistent. The Cost of Quality reports, along with other units of measure, were showing large variation from month to month and this was a cause of concern. Another issue was the high turnover rate and fluctuation in the number of temporary employees needed from day to day. The need for standardized work and standardized operator training was clear. Alan spoke with Dustin about JI and that sparked Dustin's interest in TWI as the solution to this problem.

Initial Focus on the Blow Molding Operation—The Tottle Inspection Project

Currier Plastics was experiencing costly quality problems with the new tottle process.

- The vision system was not for Tottles, did not work as planned, and was abandoned.
- The leak check systems were not for Tottles.
- There were two big problems with the tottle job. One, it was a new-style extrusion head and not too many employees understood how to troubleshoot and get to the root cause of the issues. This led to "Band-aids on top of Band-aids," which resulted in many non-conformances. Two, the inspection process was labor intensive and lacked standardized work. The first issue was addressed through the Tier 2 Technical Training program and daily talks about the process, issues, and troubleshooting. The second issue was corrected through Kaizen events, standardized work, and Job Instruction Training.

CNYTDO TWI Trainer Glen Chwala delivered additional JI training in November 2007 for classes staffed with cell leaders and lead operators from the Injection

Molding Department; the manager, engineer, and lead operators from the Blow Molding Department; and the manager, engineer and quality assurance technicians from the QA Department. Upon completion of this training, along with coaching on how to apply JI, it became clear that JI could provide a foundation to stabilizing their processes.

For the first few months every JI class attendee was involved in creating Job Instruction breakdowns (JIBs) for training, but it was not exactly clear how the TWI program was to be implemented or sustained. Unfortunately, TWI was then set aside for these people to participate in scheduled Lean events that were a priority at this stage of the Lean strategy. Rapid implementation of the improvements generated by these Lean events over several months created a whole new set of problems as people resisted change, something supervisors and team leaders were ill equipped to deal with. The leadership team responded by having CNYTDO introduce Job Relations (JR) to give team leaders and key people the skills they lacked in getting results through people.

January 2008—The Currier Strategy for Implementing TWI

A decision was made that JI would be used to train all operators on how to perform the new Tottle inspection process. This inspection process was completely different from anything that was currently being done in the Blow Molding Department, allowing operators to practice Jidoka by shutting down the blow molding machine when defects were found. Steve Valentino, the Blow Molding

Department manager, participated in the JI training and immediately saw the potential benefits of the JI standardized training. According to Dustin, "Steve became a huge advocate of the process and his support played a key role in the success of TWI in the Blow Molding Department." At this time, a technical Tier Training Program was also started that trained "newer" technicians on identifying and correcting root cause issues.

February 2008—Steve Valentino and Dustin Dreese Drive the JI Rollout

A second round of JI training was scheduled to give cell leaders, lead operators, and operators initial training to get them on a path to becoming company trainers. steve conducted meetings with the cell leaders, encouraging them to get involved in the program while Dustin worked with production, quality, maintenance, and engineering staff to create work instructions for the "future state" of the Tottle operation. They documented every task, from box making to final packaging, and then created JIBs for every step in the process.

As each new JIB was completed, it was passed around for review by lead operators on all shifts. If a lead operator approved of the JIB, he or she would sign-off and approve the JIB before returning it. Any suggested improvements by a lead person were written up on a new JIB that was again circulated to the others for review. This method played a key role in getting buy-in from all operators on all shifts. While this was going on, continuous improvement efforts continued to free up operator and lead operator time running the Tottle process by reducing cleaning time and part jams as well as by reorganizing operator workstations to create more productive workflow. These improvements were, in turn, captured on JIBs so the current best way could be used in training.

March 2008—A Lesson on Benchmarking

The timing could not have been better for Cindy Oehmigen to arrange for four people from Currier to visit the local ESCO Turbine Technologies plant that had become a national model of how to properly implement the JI program in support of standardized work.* Not only did seeing TWI in action exceed their expectations, favorable comments from the ESCO trainers, supervisors, and operators allowed the Currier visitors to walk away with advice on how to implement JI based on ESCO's experience using JI throughout their plant.

- Applying the three TWI modules of JR, JI, and JM at once is too much, too fast.
- Target a troublesome area for first application of JI.
- Train the people who will be involved in the first application of JI in both JI and JR.
- Dedicate resources for training and follow-up with reporting and conformance audits.
- Creating JI breakdowns is not as simple as it appears. Teach your people how to use JR for teams to reach consensus on the current best way.

---

* Patrick Graupp and Robert J. Wrona, *The TWI Workbook: Essential Skills for Supervisors,* Productivity Press, 2006, case study on CD that comes with the book.

- Follow faithfully the JI training method of three instructor demos and four learner trial performances, with increased levels of detail and understanding shown in each repetition.
- Identify Trained, Skilled, and Qualified levels by color-code on employee badges.
- Post the training matrix to make the status of JI training is visible.
- Provide refresher training when warranted.
- Provide follow-on support by a qualified TWI trainer immediately after JI is introduced to make sure people consistently do jobs as trained.

The four visitors from Currier walked away from their visit to ESCO with a clear understanding of the importance of follow-up coaching for a company starting out with TWI. Alan Gross immediately had CNYTDO arrange for Paul Smith to make a presentation to the Currier Leadership Team on how to set up an internal coaching and auditing process, as he had done for ESCO Turbine Technologies-Syracuse and for Albany International (Chapter 6). Paul partnered with Currier's Lean consultant Mark Gossoo of CMS Consulting, who was also a Certified TWI Trainer, to form a TWI Implementation Team. Several generic skills (often repeated operations) were identified as the first priorities for training by the team on the Tottle specific jobs. This newly formed TWI team would do report-outs every other week. Once the first group of breakdowns had been drafted, Paul and Mark were invited in to observe training using these new breakdowns at which time they coached the Lead Operators on how to use the JI Breakdowns (JIBs) to train others.

Planning and Rolling JI Out in Blow Molding

The target date of April 22, 2008 was set for all shifts to have three operators completely trained on the new Tottle JIBs for a 7 AM start-up. To make time for this training, Department Manager Steve Valentino allowed the lead operators (shift JI trainers) to shut down a "low priority" machine for 1-2 hours per shift to train the operators on the Job Instruction breakdowns. A training matrix was created as a visual control that documented who needed to be trained, on what job, and when.

Despite all the preparation, the process was new, as were the people being trained. The trainers were also new and it would be natural for them to be nervous and uncomfortable in their new role. To overcome this potential problem, which would have significant impact on production and acceptance of the new process, Paul and Mark spent one-on-one time preparing each trainer candidate by having them practice training each other and then critiquing them afterwards on how to improve. According to Steve Valentino, "Coaching by experienced trainers like Paul and Mark was the key to getting the selected trainers involved and on board with the TWI program."

April–December 2008—Launch of the New Tottle Inspection Process

At 7 AM on April 22 a large group of staff from a variety of disciplines was on hand for the launch of this new process to answer the many anticipated operator questions and to resolve issues as they came up. Initial audits identified the need for process and machine-related changes to the JIBs, and these were followed by the immediate retraining of operators. Blow Molding Department employees not trained before the launch were trained in May and immediately given opportunities to run the Tottle process. Auditing operators indicated that most operators followed the JIBs all the time while some tried cutting corners when not being observed. Operators involved in such non-conformance were retrained.

The impact of JI and standard work for the Tottle process led to creating standardized work and Job Instruction breakdowns for all blow molding jobs. As a result, 160 "part specific" work instructions (WIs) were reduced to 12 "machine specific" WIs. These have since been reduced further to 7 WIs that simplified the training in Blow Molding so that all operators could easily be trained to run all machines in blow molding, and this helped to facilitate the creation of a flexible workforce in that department. Currently that department now has a total of 28 operator JIBs in the ISO system for training on the 7 operator WIs. Alan Gross doubts the company would have gone down this path had the company not introduced the TWI JI training.

Sustaining the Training—A Challenge for Steve and Dustin

As could be expected, there were a few "renegades" who would perform their tasks according to the training when being observed, but would cut corners when no one was watching. These operators were identified through the quality assurance (QA) auditing process and retrained.

Although repeat violators appeared on the QA audit non-conforming retraining list, most operators eventually settled into doing the job as trained by the JI trainer.

Some other employees, however, could not accept change and opted to leave the company. This would have been a problem had the Blow Molding Department not been going through some rather huge changes generated by the Lean projects that enabled production to run with fewer people. By late 2008, the JI program was having a significant impact on the Blow Molding operation:

- Operator related non-conformances were drastically reduced.
- Daily temporary labor was no longer needed to sort defective product.
- A bank of eight machines that took four operators to run was reduced to two operators
- Workplace Visual Scoreboards provided employees with a sense of accomplishment while helping to drive additional efforts to improve.

Impact of JI on the Tottle Inspection Process

The "old" Tottle inspection process is referred to as 100% waterfalling. Operators would dump 1,500 to 2,000 parts from the complete box onto a table and then slowly move the parts back into the box while removing parts with visible defects. Due to time constraints and/or the quantity of defects involved, only 70 percent of the visual defects were found. As a result, unacceptable quantities of defects were being shipped to the customer while the machines that produced these defective parts continued to run during the inspection.

The "new" Tottle process requires operators to thoroughly inspect numerous points on only 10% of the parts. The QA Department monitors PPM data to determine the criteria for the inspection process. JIBs are used to train operators on the standardized way to run the inspection process, which requires the operator to shut down the machine when a defect is found, and to identify and correct the root cause of the defect before restarting the machine (Jidoka).

Sustaining the Gains and Moving Forward

Standardized work on the plant floor is now determined by a cross-functional, participative process. The TWI Job Instruction format is used extensively to identify and document "the one best way" to train employees new to the process. This process is then accelerated through a Kaizen Event style review of the proposed Job Instruction Breakdown via video tape with all shifts involved. Standardized work has also been applied to support organizations in addition to the manufacturing processes and was the basis for their ISO 9001-2008 certification. Here, process flow mapping was helpful to define the "as-is" and "to-be" states that triggered substantial improvement in the administrative functions.

A ten-hour JI training session was conducted in September, 2009 for additional cell leaders, lead operators, and operators. Operating procedures, job descriptions, and performance appraisals for everyone involved with the Currier JI Program were created and/or updated to include their JI training responsibilities and then added to the ISO System to help sustain the program in the Blow Molding Department.

Another ten-hour JI class was conducted for engineers, maintenance, cell leaders, operators, lead operators, and office personnel prior to the launch of the second phase to expand standardized work and training for positions beyond the operator function in the Blow Molding Department. Dustin initiated monthly meetings with the lead operators to create a "back-up" lead operator training program. They standardized work for the lead operator position and created JIBs to train a "back-up lead operator" on each shift to cover vacations, call-offs, or to fill openings. The success of this approach led Dustin and Steve to extend the JI training by creating standardized work for other job classifications in Blow Molding to create a highly flexible workforce capable of performing many different jobs in that department on any given day. Alan was now ready to take all this learning to Injection Molding.

Moving On to Injection Molding

It was now obvious to Alan Gross that TWI and Lean resources were needed in Injection Molding to accomplish what Dustin and his people had accomplished in Blow Molding. With Cindy's help at CNYTDO, Alan found the right person for the job.

After receiving her BS in ceramic engineering from Alfred University, Jennifer Dietter advanced from lab technician at General Electric to ceramic engineer at

General Color and Chemical in Ohio before joining Syracuse China in 1996. Her work in developing trouble-shooting guides at Syracuse China to reduce defects led Jennifer to create Standard Operating Procedures and training classes she conducted on the operation of a new glazing machine. The success of this training led to the creation of an internal training department in 2004 that Jennifer led and she became the Lean Facilitator when Lean was introduced in 2006. Jennifer also became a Certified TWI trainer for the "J" programs in 2007, making her the ideal candidate to take on Dustin's role in Injection Molding as a dedicated consultant when Syracuse China was closed by its parent company, Libbey Glass, Inc.

Dustin and Jennifer started working together as a team when she came on board in April 2009 to drive the development of work instructions in the Injection Molding Department. As the internal resource for TWI training, Jennifer was quickly put to work delivering the ten-hour JI classes. Monthly meetings were set up for Dustin and Jennifer to work with Lead Operators to update Work Instructions and to create JIBs as needed. This was a major task since most of the JIBs they were creating for Injection Molding were item specific, as opposed to being process specific in Blow Molding. By the end of the year Jennifer had taken over Dustin's role leading the monthly meetings, along with the coaching and auditing of Lead Operators using JI. This freed up Dustin's time for other responsibilities.

Jennifer also introduced JM to Currier Plastics by delivering several ten-hour JM training classes to those with continuous improvement as a part of their daily responsibilities. These classes led to the development of an extremely effective kanban system in Blow Molding, and, in Injection Molding, TPM and an improved Maintenance Request Process resulting in a more visual maintenance process. Improvements made through use of JM are posted on a JM Education Board located on the production floor. Additionally, Jennifer will be delivering the first two hours of both the JR and JM ten-hour classes to introduce JR and JM to everyone in the company. This will allow everyone to use the same language and understand expectations when dealing with change regarding people and evaluating improvement ideas.

Keeping Score with Lean Accounting

Throughout this process, Currier was moving forward on various initiatives as outlined in its Lean strategy, including Lean Accounting, Supply Chain Management, and other Quality initiatives. The company is now benchmarked as best in class for its Lean accounting approach. Each of its two major value streams has weekly Box Scores communicated throughout the value-steam. Each organization's Box Score includes a few vital signs that are leading indicators of operational performance. Because the system provides them with independence and accountability, each value stream has the resources it needs to run its business, relying little on central support services. Its Box Score facilitates quick decision making, keeping their overhead expenses minimal.

Operationally, visual control boards are located in each department, like the one shown in Figure 11.7, which identifies waste in the injection molding area. Each area

Figure 11.7 Injection molding visual control area board. (Reprinted with permission of Currier Plastics, Inc., Auburn, NY.)

board has all measures linked to the strategic objectives for that area as discussed earlier in this chapter. Reports are color coded to represent performance to the benchmarked goals established and updated quarterly. Beyond day-to-day management, this helps focus special attention on areas for improvement signaling the need for a kaizen or Six-Sigma event.

Supply Chain Management

Currier defines its supply chain as "from dirt to dirt," from the source of raw materials used to make resin to when its products are returned from use to a landfill or reused. As a result, the Currier supply chain management strategy is heavily influenced by both Lean thinking and sustainability objectives.

From a Lean perspective, the company has worked with their significant suppliers and customers to establish common inventory goals where work in process (WIP) and finished goods are visible to both parties. Transportation costs have been reduced in spite of adding more frequent deliveries as the company reduced internal batch quantities and setup times and worked with partners putting into place kanban and returnable packaging programs. From a sustainability point of view, departments have reduced scrap and increased internal recycling, thereby removing all plastic from the disposal streams. Where practical they have specified use of post-consumer and industrial recycled materials and, for the first time in the United States, sampled biodegradable and natural fiber-rich resins. Numerous proposals have been made to customers to eliminate non-value-added packaging.

Quality

Quality improvement takes many forms at Currier Plastics, where its corrective action process incorporates customer complaints, internal systemic issues, continuous improvement and preventative projects, safety issues, and internal audit findings using a format that mimics the Six-Sigma DMAIC process. A cross-functional Quality Review Council (QRC) meets weekly to review progress and provide direction to corrective action/project leaders. "Tips of the day" are circulated via e-mails and postings alerting plant-wide personnel to recent findings and learning.

The Quality Assurance group is actively involved in all new product development activities. Their engineering function includes a fully functional lab with the latest testing equipment, including a fully programmable vision/touch probe inspection device, a tensile/compression tester, microscopic measurement devices, a vacuum test chamber for leak testing, torque tester for bottle/caps, and a moisture analyzer for resin testing. While the majority of production inspection is done on the floor, quality technicians provide more advanced testing and Statistical Process Control (SPC) functions.

SPC on critical dimensions is done by direct gauge entry on the floor and control charts that automatically signal concerns and out-of-control situations and notification to QA. Quality is further assured using the SPC capabilities for key molding process parameters (i.e., times, temperatures, and pressures) to prevent suspect product being made in the first place. These may signal that a machine or tool is in need of maintenance by identifying a wear condition that could affect the process.

Indicators of Success

Performance trends are displayed using Four Up Charts for each value stream that track the most vital of the Key Results Measures for operations: productive downtime, overall efficiency, quality, and delivery performance. These are posted on the shop floor Lean bulletin boards and updated weekly, acting as the Box Score Summary for production teams to monitor their performance. As can be seen from these charts (see Figures 11.8, 11.9, 11.10 and 11.11), the Lean strategy has paid off handsomely.

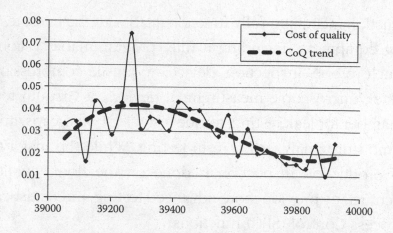

Figure 11.8   Cost of quality as a percentage of sales. (Reprinted with permission of Currier Plastics, Inc., Auburn, NY.)

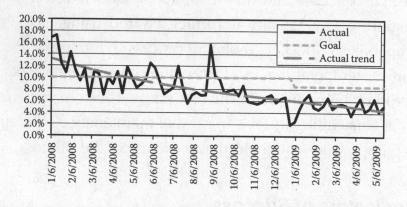

Figure 11.9   Productive downtime. (Reprinted with permission of Currier Plastics, Inc., Auburn, NY.)

Closing Remarks from Vice President of Operations Alan Gross

Changing the culture is never easy. It takes persistent and consistent time, involvement, leadership, and energy. It's not enough that some are advocates and others are fence sitters. All key personnel need to lead. We are creating a culture of continuous improvement in literally

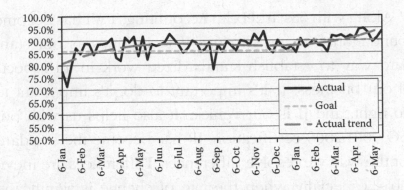

Figure 11.10 Overall efficiency. (Reprinted with permission of Currier Plastics, Inc., Auburn, NY.)

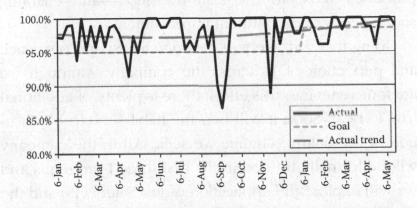

Figure 11.11 Ontime delivery. (Reprinted with permission of Currier Plastics, Inc., Auburn, NY.)

everything we do—that's what V^2 represents. For us at Currier Plastics, this transition included a recognition that a changing skill set was needed to take us where we wanted to go. It also meant less people were needed. Although this transition was difficult, we have become a much stronger company as a result. Part of the culture change is the recognition that open, honest, no-holds-barred communication was needed in all teams—in fact, in all meetings.

Along with 6Ss and Lean Accounting, TWI has become foundational for us. We use JI techniques as our standard way to establish standardized work in all aspects of our business. If it's important to do, it's important to do right, and JI is our vehicle. It also helps us with our ISO 9001:2008 certification. JR has become the standard for the way we resolve differences. Differences are inevitable, especially when the rate of change is significant, like at Currier Plastics. JM, which has just been introduced, is quickly becoming our standard for improving processes—core to the lean transition—and enabling spontaneous and continual improvement.

Going forward, we will continue to expand our reach and perfection of JI across the company. Although we are four years into this effort, there is plenty of additional fruit to bear. Soon it will become habit—the true indication of the culture change we seek. All in the company will be formally trained in the first three J-programs and we will explore the application of Job Safety (JS) and the TWI Problem Solving (PS) methodology taught by the TWI Institute.

V

EXPANDING TWI BY LEVERAGING JR, JI, AND JM

Chapter 12

TWI's Problem Solving Training

In the spring of 1951, Lowell Mellen and his associates from TWI, Inc. began teaching TWI classes in Japan under contract with the U.S. military occupation. Mellen had been a district representative of the TWI Service in Cleveland during the war and he formed TWI, Inc. in order to continue implementing the program when the service disbanded in September 1945 at the end of WWII. After successfully planting the three original J programs in Japan, TWI, Inc. was asked by the Japanese government in 1955 if they could teach supervisors how to solve workplace problems, and building on the success TWI was enjoying in Japan to that point, they then developed a new TWI program called Problem Solving Training.

It seems strange today to think that the Japanese, with their manufacturing prowess in resolving production problems that results in the exceptional quality and durability of the goods they produce, would need lessons in problem solving. But things were different during these early years when TWI was still a new program. In a 1956 report, TWI, Inc. described how Japanese industry was controlled by "old hands" who didn't want to give up their "feudal prerogatives," and managed not by "definite policies, principles or fundamentals but by whim and caprice":

> This means that at all Supervisory levels below the very Top there is a reluctance to accept Responsibility for anything ... so there is a constant crisis in waiting for someone to make a decision or come up with the answer to anything. The final result is that the average Japanese Supervisor is a very, very frustrated individual who hardly knows which way to turn.*

The "average Japanese supervisor" has certainly come a long way since then! By the time Patrick arrived in Japan in late 1980, front-line supervisors at Sanyo Electric Co., Ltd., where he worked, knew exactly which turns to take and were active in confronting and solving the difficulties they faced each day. And the TWI Problem Solving Training was a key piece of their training regimen.

To keep things in perspective, compare the situation in Japan in 1956 to the situation found at GM in 2009 after the company went into bankruptcy and was forced to open itself up to outside review and evaluation:

> When G.M. collapsed last year and turned to the government for an emergency bailout, its century-old way of conducting business was laid bare, with all its flaws in plain sight. Decisions were made, *if at all* [emphasis added], at a glacial pace, bogged down by endless committees, reports and reviews that astonished members of President Obama's auto task force.†

---

* TWI, Inc., Final Report, 1956, p. 32.
† After Bankruptcy, G.M. Struggles to Shed a Legendary Bureaucracy, *New York Times*, November, 12, 2009.

While GM certainly learned a great deal about Lean and other quality techniques from its twenty-five-year relationship working directly with Toyota in their joint venture plant in Fremont, California, they never updated their command-and-control corporate culture to a more humanistic approach, which is the essential change that needs to take place for the Lean tools to work (see Chapter 7). As we pointed out in the first chapter of this book, the lessons learned in the U.S. during World War II developing and using TWI were lost after all the GIs came home and took their old jobs back, doing them the same way they did before they left for the war. In Japan, though, as we will now see, they not only fully embraced the TWI methods, but expanded on them and pushed problem solving and decision making down to the factory floor.

Comparing TWI and Toyota Problem-Solving Methods

Toyota is well known for, among many other things, its effective approach to problem solving with a focus on finding the root cause. So let's start by comparing the Toyota problem-solving steps to the TWI method to show the direct correlation between the two. Knowing the strong influence TWI had on the early development, and current foundation, of the Toyota Production System (TPS), it will not be surprising to see how the two problem-solving approaches neatly align.

At the 2009 TWI Summit held in Cincinnati, David Meier, coauthor of *Toyota Talent*, outlined the eight steps

Table 12.1 Comparison of Toyota and TWI Problem-Solving Steps

Toyota Problem-Solving Steps	TWI Problem-Solving Steps
1. Clarify the problem	1. Isolate the problem
2. Break down the problem	• State the problem
3. Target setting	• Give proof or evidence
4. Root cause analysis	• Explore the cause
5. Develop countermeasures	• Draw conclusions
6. See countermeasures through	2. Prepare for solution
7. Monitor both results and process	• Use JM, JI, and JR steps 1 and 2
8. Standardize successful processes	3. Correct the problem
	• Use JM, JI, and JR steps 3 and 4
	4. Check and evaluate results

to problem solving currently being taught at Toyota. Table 12.1 shows the Toyota problem-solving steps side by side with the four-step method of TWI Problem Solving developed in 1956 by TWI, Inc.

Both of these approaches spend a lot of effort at the beginning of the process seeking to correctly define the problem before beginning to solve it. As the saying goes, "A problem well stated is a problem half solved." Meier stressed in his presentation that, when dealing with a problem, where most people go wrong is that they immediately go out and try to fix it before they know what it is that is really broken. In other words, they want to go straight to the answer without taking the time to be sure they are dealing with the right question. The result of this haste and lack of clarity is to take shots at fixing symptoms, or perhaps even the wrong problem itself, while the true causes go on and continue to generate headaches.

Step 1 of TWI Problem Solving in effect covers the first four steps of the Toyota method. In both cases, the key is to get into the details of the problem, found at the *gemba*, so that we can understand clearly what we are actually dealing with. When we break down the problem, we are looking for the underlying components that are generating the visible problem, the thing that is causing the pain, and we can best do that by finding the proof or evidence of the various factors feeding the problem. In this way we'll know that the problem exists as stated and have some idea of where it is coming from and on what scale. In other words, we should not jump to solutions after a cursory review of what we see happening on the surface of a problem.

Once we understand the details of the problem, then we can proceed to the ultimate goal, which is to determine the root cause of the problem. In TWI, when we say "explore the cause," we mean just that, finding the core or root cause, which the TWI developers called the problem point. As we shall see, the TWI program devised a method of "digging down deep" to find the root cause of the problem, and this tool looks identical in practice to the famous Toyota method of asking the "five whys." It is only when we get to the true cause of the problem that we can successfully correct it.

Where the two methods diverge is in how they prescribe resolving the problem at hand. The Toyota plan is much more open-ended, using tools such as brainstorming and consensus building in order to develop and enforce countermeasures. It stresses the vital need of taking time to work with other people related to the process in order to get their buy-in to the correction process. The key here, as John Shook explains in *Managing to Learn*, is to develop "a set of potential countermeasures rather than just one approach."[*] This allows for risk to be minimized by reviewing a variety of scenarios, and it also, more importantly, allows other stakeholders to take a more active role in evaluating and fine-tuning the correct solution. With various possible solutions on the table, the supervisor can entertain the needs and concerns of others while creating consensus and support around attaining the desired outcomes.

While the Toyota method takes this high-level approach to finding solutions, the TWI method gives specific prescriptions for solving the problem: the use of the TWI methods of Job Methods Improvement, Job Instruction, and Job Relations. It analyzes the core cause to determine if the problem is mechanical or people, or both, and then applies the proper TWI tool to bring that problem to a solution. Though this approach is narrower in scope than at Toyota, it allows front-line supervisors to use the TWI skills they possess, which are easy to learn and apply, quickly and effectively to deal with the problems they face on a regular basis.

Finally, both the TWI and Toyota methods finish by thoroughly monitoring and evaluating the results of the problem-solving effort in a way that sustains the corrected processes. The Toyota method focuses a bit more on standardizing and sharing successful processes that have

---

[*] John Shook, *Managing to Learn: Using the A3 Management Process to Solve Problems, Gain Agreement, Mentor and Lead*, Lean Enterprise Institute, Cambridge, MA, 2008, p. 75.

been found, an aspect covered in the Job Instruction portion of TWI, while the TWI problem-solving method puts more emphasis on monitoring the people's reaction to the change—people resist change and see it as a threat to their basic needs—and looking for signs of new problems being created by the correction. But these aspects are well covered in both the Toyota and TWI methodologies.

It appears evident that the straightforward and commonsense approaches to supervisory management techniques originally proposed by the TWI programs in the 1950s are clearly reflected in the successful Toyota approaches to everything from standardized work to problem solving. The main differences, we feel, in these two methodologies are the addition to the Toyota method of key aspects of Japanese culture and practice, such as consensus building, called *nemawashi* in Japanese, a focus on human development and mentoring, and the use of concise, easy-to-read and -understand tools like A3 reports to support effective communication. These techniques are extremely powerful in bringing about lasting improvements, but they nonetheless do not take away the need for strong skills, as represented in the three basic TWI methods, in taking effective countermeasures to the problems themselves.

Therefore, the two approaches are quite complementary and can, and should, be used interactively. The A3 management process is, in many ways, a philosophical approach to dealing with the people who will be dealing with the problem. The TWI method is more of an "on the ground" tactical manual for front-line supervisors in solving problems. There are many good works describing the Toyota methodology, so here let's take a more in-depth view of the TWI problem-solving method.

What Is a Problem?

"A supervisor has a problem when the work assigned fails to produce the expected results." This straightforward definition of a problem is as true today as it was when it was written over fifty years ago in the original TWI Problem Solving manual. In that manual they created a simple graphic to display this gap between the current situation and the needed standard (see Figure 12.1). Today, this model has been modified to reflect what is called the gap analysis, but as can be seen in Figure 12.1, the idea is the same.

Some of the problems supervisors face are small and fairly simple and can be solved quickly by relying on the supervisor's experience and good judgment. But

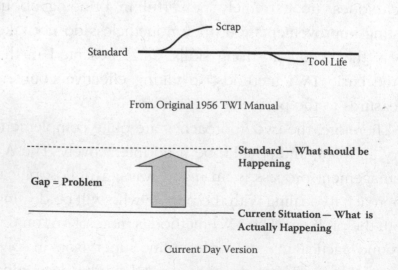

Figure 12.1   Gap analysis—old and new.

most of the problems we face will ultimately turn out to be much more complex and important than we at first anticipate, and we need to have an organized method of analyzing them before we rush to a decision on what to do. In order for supervisors to handle problems successfully, then, they need a clear sense of direction and specific steps that will lead them to a successful solution.

The TWI Problem Solving manual specified four facts that supervisors must understand if they are to resolve production problems successfully:

1. Their problems are seriously interfering with their ability to get out production.
2. Their problems have a direct effect upon cost and quality.
3. Their problems are not necessarily inevitable.
4. Their problems can be solved easily by following a definite pattern.

The first two points have to do with an awareness of one's own responsibility toward resolving problems. Even as supervisors struggle daily to get out quality production while keeping costs under control, it's easy to be in denial about the problems surrounding them. They may consider these problems to be mere nuisances that get in their way, and they don't realize that their struggles emanate from the very conditions they are trying to ignore. The second two points indicate that, once they realize that they have problems that need to be addressed, there is a way to handle these problems.

Without a good method to handle problems, supervisors may have difficulty determining just what the problem is and confuse evidence of the problem with the problem itself. For example, if a machine is running poorly because it is low on oil, once the supervisor finds this out, he refills the oil and believes he has solved the problem. But the low oil level is just a symptom of a leaky valve, and so the supervisor's action is not conclusive and the problem continues, perhaps with even more serious consequences. Unless we are able to find the true cause of the poorly running machine, we will not be able to take the action that brings it back to normal performance.

Before we take action on any problem then, the first and most important step is to find, or isolate, the true problem.

Step 1: Isolate the Problem

As we pointed out earlier when comparing the TWI and Toyota methods, this first step of TWI problem solving consists of four distinct activities: (1) state the problem, (2) give proof or evidence, (3) explore the cause, and (4) draw conclusions. In this step we want to clearly define the parameters of the problem so that we can search for the root cause. Once we find the root cause, we can decide on a solution course based on the makeup of the problem. Let's take a look at each one of these points.

State the Problem

When a problem is large and significant, it is not difficult to state what it is. But typically, problems are ongoing and in a continual state of flux. So they might be difficult to pin down. For example, you might find that you're having difficulties keeping up with the production schedule, or that the morale of the workforce is not what it should be. How would you state these so that you knew more specifically the problem being addressed? We'll get into the actual figures, the data, of the problem in the next part, "Give Proof or Evidence of the Problem." But for now, to be more clear, you could state, "parts delayed coming out of final inspection," or "many minor errors due to careless mistakes." Careful consideration in defining just what it is we're trying to attack, rather than vague generalities or mere complaints, is the way to get off to a good start.

This does not mean, though, that we have to restrain our efforts to problems that are happening right now. We can also look for new problems just breaking out or ones that we see approaching so that we can get into them early, while they're still small and easy to handle. We can also seek out and find problems by talking with other people and getting their perspectives and opinions, by reviewing reports and records to discover trends or hidden issues, or by simply anticipating problems and using our intuition. Having a "problem consciousness," or *mondai-ishiki*, as the Japanese call it, means to always be aware and sensitive to where problems might be lurking so that we can take problem-solving actions to neutralize them before they interfere with our work.

Once we have identified the variety of current and potential problems we are facing, we should then prioritize them so they can be tackled one by one in order of importance. We can prioritize them by the level of impact they will have—some problems may be deemed emergencies that need immediate attention, while others might be considered threats that should be taken care of just as soon as reasonably possible. We could look at them by their level of importance or necessity, in other words, by the amount of benefits we receive if they are solved. Or we could look at them by the level of difficulty it would take to investigate and resolve them to see what amount of investment is necessary and available to resolve them. Based on this assessment, we can allocate our resources in the most effective way.

Give Proof or Evidence of the Problem

By proof or evidence of the problem we mean the actual facts, circumstances, figures, etc., that directly show the extent of the problem and prove that it is, in fact, something that needs to be dealt with. It is important at this early stage to look at data, wherever appropriate, that is concrete, well confirmed, and to the point. When faced with a problem, it is easy to get flustered and chase after something we think is causing trouble before we have confirmed its true validity. As we stated earlier in this chapter, we should resist the temptation to jump to solutions. On the contrary, once we have gotten specific data around a clearly identified problem, we are in a much better position to begin searching for the causes of that problem because we have taken the first step to locating just where those causes may be lurking and to what degree they are afflicting us. We are not on a wild goose chase—we have the proof and evidence.

What kind of evidence should we be looking for? This may at first appear self-evident: get data that are directly affecting the problem. But the TWI method makes a breakthrough here by beginning an analysis technique that continues throughout the entire problem-solving process: looking at the situation from two angles—the *mechanical* and the *people*. Especially in manufacturing settings, we commonly look at most everything from the aspect of physical things, like machines, parts and materials, tools and equipment, handling and storage, designs, processes, facilities, and so on. But in doing effective problem solving, what we find, more oftentimes than not, is that what begins as a straightforward mechanical problem turns into, after careful analysis, a fundamental problem concerning people. As we point out in Job Relations, people are at the center of everything we do in fulfilling our responsibilities. So we must look equally at both aspects, the mechanical and the people.

From the mechanical angle, we look at things like:

- Schedules
- Rework and scrap
- Tool wear and breakage
- Equipment breakdown
- Accidents
- Setup times
- Records and paperwork

From the people angle, we look at things like:

- Productivity
- Work habits

- Knowledge and skill
- Safety
- Responsibilities
- Attitude and interest
- Job satisfaction
- Personality
- Physical condition and health

We have to take a fresh and honest approach when looking for this proof and evidence. In other words, we have to address the problem with an open mind and let go of our preconceived notions around the situation. We can do this by talking with people related to the problem to get new insights or contrasting points of view. In particular, if there are people who are directly involved in the problem, we need to find out who they are so that we can get a firsthand account of the facts. The purpose here, of course, is not to lay blame, but to be sure we understand the true details of the situation. We should also review any available records to see if there is a history of similar or related issues.

Once we get these data, we should record them as concretely as possible. The TWI Problem Solving program created an easy-to-use problem analysis sheet to facilitate this (see Figure 12.2). The facts from this part of the process are recorded under the section for proof or evidence, right under the statement of the problem.

Explore the Cause

Once we have identified visible evidence of the problem, the next thing to do is to find the causes, both direct

Problem Analysis Sheet

Problem		Name: _____
		Date: _____

	Mechanical Angle			People Angle		
Proof or Evidence	Behind schedule by			Productivity is		
	Rework is up by			Work habits are		
	Scrap is up by			Job interest is		
	Tool breakage is up by			General attitude is		
	Machine time is down by			Work quality is		
	Accident rate is up by			Complaints are		
	Setup time is			Attendance is		
	Paperwork increased by			Job satisfaction is		

Why?   Where?   When?   Who is responsible?

	The Above Problems Are Caused By:			The Above Problems Are Caused By:		
Causes	Job method			Incorrect job assignment		
	Layout			Insufficient skill and experience		
	Tools, fixtures, gauges, etc.			Faulty instruction and follow-up		
	Machines and equipment			Poor human relations		
	Materials and parts			Personality situation		
	Product design			Basic wants threatened		
	Housekeeping and working conditions			Health and physical fitness		
	Unsafe conditions			Unsafe acts		

This problem concerns: Quality? ○ Cost? ○ Quantity? ○ Safety? ○ People? ○

	Problem Points (Core/Root Causes)	Mechanical	People	
Conclusion		Things/Places	Don't Know Can't do	Don't Care Won't do

Figure 12.2 Problem analysis sheet.

and indirect, of each and every piece of evidence. When we search for these causes, we have to throw away our subjectivity and prejudice and think broadly. As the TWI method teaches, we have to dig down deep in order to get to the source of the problem, looking at each piece of evidence and asking why, where, when, and whose responsibility. In doing so, we seek to get to the problem point, or the root cause.

Here again, the method looks at the two aspects of *mechanical* and *people* when exploring the cause of each piece of evidence. While the topics are the same, what we are looking at now are those specific aspects of the process that contain the reasons for the problem.

From the mechanical angle, we look for causes in:

- Method
- Layout
- Tools and equipment
- Materials
- Design
- Environment
- Unsafe conditions
- Standards

From the people angle, we look for causes in:

- Job assignment
- Insufficient knowledge/skill
- Faulty instruction
- Human relations
- Personality/character
- Unsafe acts

We record these causes in the appropriate section of the problem analysis sheet right under the proof or evidence (see Figure 12.2). In looking over these characteristics, we must consider the facts and piece together the cause-and-effect relationships between the various causes. In other words, we have to understand how there are causes of causes so that if we keep drilling down, we ultimately find the problem point, or root cause. This analysis leads us directly into the last section of Step 1, draw conclusions, where we select out these core causes as targets for our countermeasures (see Figure 12.2).

The original TWI problem-solving plan outlined this part of the method as we have described here. However, the Japanese later felt that the analysis was a bit lacking in that, while it sought the root cause, it did not lead us there in a direct path. So they created another analysis form to be used at this point, which better captured the cause-and-effect relationships we are seeking. By listing the problem, evidence, and causes—direct, indirect, and core—in a chain of causation, we can immediately see how these factors are all related to each other. So we call it the chain of causation analysis sheet, and we can see an example of its use in Figure 12.3.

Now that we have lined up this chain of causes to the stated problem, it's not hard to see that what we are doing here is continually asking why at each stage of discovery until we get to the bottom of the problem, or the root cause. This is exactly what Toyota does in its root cause analysis, what has been called the five whys, or asking why five times until we get to the root cause. Looking at Figure 12.4, we can see that that is exactly what we're doing with the chain of causation analysis,

Problem	Evidence	Causes		
		Direct	Indirect	Core/Root
Customers complaining about late deliveries	On-time delivery stands at 87%	Packaging delays creating a bottleneck of finished product	Delivery of packaging product is frequently delayed from printer Films from art department are being held up waiting for confirmations Need corporate approvals for correct usage of all company logo marks	Poor communication between corporate marketing and plants

Figure 12.3   Chain of causation analysis sheet: Example.

noting that, under indirect causes, there continue to be causes of causes until we finally find the core cause.

Draw Conclusions

The root cause analysis we have just completed does a good job of helping us identify the problem points. However, before we move on to try to solve them, the last thing to do in Step 1 is to draw conclusions from our analysis and decide on a solution course. We need to ask ourselves, "If these causes were removed, would we still see evidence of the problem?" In other words, have we indeed found the true cause of the problem that is generating these difficulties? Is this a problem in my own area of responsibility? Should I handle it myself

Problem	Evidence	Causes		
		Direct	Indirect	Core/Root
Customers complaining about late deliveries.	On-time delivery stands at 87%.	Packaging delays creating a bottle-neck of finished product.	Delivery of packaging product is frequently delayed from printer. Films from art department are being held up waiting for confirmations. Need corporate approvals for correct usage of all company logo marks.	Poor communication between corporate marketing and plants.

Figure 12.4   Chain of causation analysis sheet: Example showing "5 Whys."

or report it to someone else? If it is my responsibility to handle, should I delegate it to someone below me or get help from other specialty staff functions, like engineering, quality, costing, etc.?

Earlier in this step, we saw how we began to distinguish between the mechanical and the people aspects of the problem. Here, then, we have to determine if the problem points found are mechanical, people, or both. This determination will give us our solution course based on our TWI skills of Job Methods Improvement, Job Instruction, and Job Relations. If the problem is *mechanical* in nature, if it concerns things, places, methods, designs, etc., then we will use our JM skill, along with other tools, to pursue a solution. If it is *people* in nature, then we have to think a little more deeply about the nature of the problem.

When problems are centered on people, we need to think about what is generating the problem behavior. Human motivation is complex, but we can do a simple analysis to help guide us to effective solutions. Since our focus is on resolving specific problems that have a bad effect on supervisors' duties to produce quality products or services at the proper cost, we can look at how people and the work they do are interfering with that goal. If the problem is because they don't know or can't do the tasks assigned to them, then the solution course is to use our Job Instruction skills. If the behavior is because they don't care or won't do those tasks, then the problem is a personal situation and we use our Job Relation skills to solve it (see Table 12.2).

When deciding on any of these courses, we need to always keep in mind the company policies as well as the policies and direction of our own superiors. Even as we move forward toward solving problems in our own areas, we need to keep those actions in alignment with the overall goals and values of the organization. In describing how Toyota begins the process of problem solving, David Meier stressed the need for holding onto the long-term vision that will guide all of our

Table 12.2 Determine a Solution Course Using TWI Skills

Mechanical Problems:
• Use JM and other tools
People Problems:
• Don't know/can't do—use JI
• Don't care/won't do—use JR

decisions. This means adhering to our purpose, values, and philosophies (beliefs) to guide our strategic execution. This is especially true when it comes to taking actions on solving problems since these daily actions reflect the culture of the company we seek to build and support.

At this point we also need to consider how much of the problem we will take on at this time. Should we consider looking for a broad solution or stop at a partial solution? The important thing to consider here is which course will move us forward so that we can make progress on solving the problem. If we initially bite off too much, we quickly become bogged down and discouraged. The smarter course would be to address immediate concerns up front and then go on to address other problem areas later.

Summary of Step 1

Step 1 is the most important part of the TWI problem-solving method because it systematically and precisely identifies the correct areas to address when going about solving a problem. By stating the problem clearly, giving proof or evidence of the problem from both the mechanical and people angles, then exploring the causes, both direct and indirect, of each and every piece of evidence, we can get down to the specific problem points that need to be addressed by our solution plan. Most significantly, because we have analyzed the problem from the mechanical and people angles, we can now determine the best course of action based on the nature of these root causes. Namely, we apply appropriately our TWI skills of Job Methods Improvement, Job Instruction, and Job Relations to correct the problem.

Step 2: Prepare for Solution

Simply put, Steps 2 and 3 of the TWI Problem Solving method are a recap of the four-step methods for each of the three TWI programs: JM, JI, and JR. In other words, we implement these three TWI methods when we prepare for solution in Step 2 and when we correct the problem in Step 3. Step 2, Prepare for Solution, takes the preparatory phases of each of the TWI methods and applies them to the appropriate problem type (see Table 12.2). Then, Step 3, Correct the Problem, completes the remainder of each method so as to bring the problem to solution. Finally, Step 4, Check and Evaluate Results, is the follow-up step that ensures the problem is solved and does not reoccur. It is beyond the scope of this book to explain the TWI methods of JM, JI, and JR in any level of detail; however, a complete description of each method, along with implementation examples, can be found in *The TWI Workbook*.

Here, let's get a general overview of each problem-solving step, including a few additional tools and insights that were not part of the original TWI J methods. Figure 12.5 shows the details of Step 2, Prepare for Solution, and readers familiar with the TWI methods will recognize that the contents here, for the most part, come directly from the four-step methods of JM, JI, and JR.

Determine Objective

Mechanical Problem	People Problem	
	Don't know / Can't do	Don't care / Won't do
Job method-layout-tools-materials-equipment-design-environment		Attitude and behavior correction
Method improvement	Knowledge and skill development	Attitude and behavior correction
Analyze	**Get ready to instruct**	**Get the facts**
1. Overall situation	1. Prepare yourself	Review the record
Flowchart	Make a plan for training	Find out what rules and customs apply
Flow diagram	Break down job for instruction	Talk with individuals concerned
Question overall job	List important steps	Get opinions and feelings
2. Specific situation	List key points	Be sure you have the whole story
Word method—Method breakdown	2. Prepare the workplace	**Weigh and decide**
Layout—Discuss with operators	Correct equipment, tools, and materials	Fit the facts together
Question every detail	Have workplace set up properly	Consider their bearings on each other
Why is it necessary?	3. Prepare the learner	What possible actions are there?
What is its purpose?	Put the person at ease	Check practices and policies
Where should it be done?	State the job	Consider objective and effect on individual, group, and production
When should it be done?	Find out what the person already knows	Don't jump to conclusions
Who is best qualified to do it?	Get the person interested in learning the job	
How is the best way to do it?	Explain tools, equipment, and safety gear	
	Place the person in the correct position	

Figure 12.5   Step 2: Prepare for Solution.

Prepare for Solution of Mechanical Problems: Overall Situation Analysis

We said that mechanical problems are those involving things like methods, layout, tools, equipment, materials, machines, etc. While it is certainly true that all production problems will involve people in one way or another, either directly or indirectly, the fact that we are dealing with mechanical processes in producing a tangible product or service means that there will be many problems that are mechanical in nature. By mechanical we mean not only the use of industrial equipment, but any method or process of getting work done, including the handling of paper forms, the use of computers to process data, the creation and delivery of a service, medical procedures, and so on. These will affect the work in critical ways, such as:

- Quality—Scrap, rework, spoilage, etc.
- Quantity—Schedule bottlenecks, etc.
- Safety—Accidents, unsafe conditions and situations, etc.
- Cost—All of the above factors increase costs.

The first thing we want to do for mechanical problems is to *question the job as a whole* in order to bring to light troublesome or costly factors, such as transportation, inspections, delays, material handling, safety hazards, etc. Once we have found out in Step 1 the problem points, or root causes, of the problem, the goal in Step 2 is to search for the answers that eliminate the causes to these problems. This initial broad view of the work will oftentimes reveal the needed answers. For example, if our root cause analysis showed that the problem point was too much scrap caused by excessive handling of the raw materials, an analysis of the overall process would bring out where, when, and by whom the materials were overhandled and suggest improvements on how this handling could be reduced to prevent the scrap.

When we question the work we have to keep an open mind. As the TWI developers pointed out, "The mind is like a parachute—it functions only when open." If we are going to find openings and opportunities for improvement and for solving problems, it is absolutely necessary to cultivate an open mind. To do this we must develop and maintain a questioning attitude. In other words, we need to question everything and never accept any method or procedure as being perfect. The greatest obstacle we have in problem solving is not created by technical difficulties, but rather by the mental road blocks we set up for ourselves when we cling to the attitude that we already are using the best methods available.

The TWI Problem Solving course pointed out three conditions that supervisors need to heed if they are to move on to effectively solving a problem with a specific job:

1. Just because a job is done in a certain way is no proof it is the best way.
2. Just because a job is done at all is no proof that it is necessary.
3. The fact that the method of doing the job has been in effect for years is no proof that it is the best way.

There are many tools available to us today to make this kind of a broad analysis. The original TWI Problem Solving manual taught the use of the flowchart, which, at a glance, allows us to see the routing and the production flow of the subject being analyzed (see Figure 12.6). The subject of analysis, which we have identified in Step 1 as a problem point, can be a part, assembly, material, paperwork, etc., but usually it is a part, so that we can use the flowchart to resolve problems with work methods, layout, tools, materials, equipment, and so forth. The flowchart shows us the relationship between the subject and prior and subsequent operations, handlings, inspections, and storages. By clarifying the bigger picture of the problem area, we want to "cast a wide net" in our search for the correct solution so that we don't miss critical areas that have an influence on the problem.

A second tool the TWI method introduced for this purpose was the flow diagram, which was none other than a graphic picture, or map, of the area to show the movement of the subject using lines and symbols (see Figure 12.7). This is very similar in many ways to what we practice in Lean as a standard worksheet. It gives a more graphic view of the analysis area and makes it easy to see the layout of the process in order to make plans, reduce travel and movement, eliminate wasted space, etc.

There are yet other, more contemporary, Lean tools, such as value stream maps and standard work combination sheets, which can be used here for the purpose of viewing the job as a whole in order to discover what

FLOW CHART (EXAMPLE)

Subject Charted

Part: Angle Plate _____

Material:_____

Person:_____

Paper Form:_____

Date:_____

Department:_____

Charted by :_____

	What and How Current–Proposed Method	Oper. No.	Dept.	Dis- tance	Time	Symbols ⇨	D	○	□	▽
1	Materials Warehouse					•	•	•	•	•
2	Move to Press Workshop			240 ft.		•	•	•	•	•
3	Cut and Stamp					•	•	•	•	•
4	Move to Boring Section			90 ft.		•	•	•	•	•
5	Cut Hole					•	•	•	•	•
6	Move to Temporary Storage			30 ft.		•	•	•	•	•
7	Wait for Inspection				25 min.	•	•	•	•	•
8	Inspect					•	•	•	•	•
9	Move to Bolting Area			60 ft.		•	•	•	•	•
10	Bolt on					•	•	•	•	•
11	Move to Finished Parts Warehouse			180 ft.		•	•	•	•	•
12	Finished Parts Warehouse					•	•	•	•	•
13						•	•	•	•	•
14						•	•	•	•	•
15						•	•	•	•	•
16						•	•	•	•	•
17						•	•	•	•	•
18						•	•	•	•	•

Figure 12.6   Flow chart: Example.

Flow Diagram

Process name	Model number	Model name

Scope of Operations
From:
To:

Date:
Name:
Department:

Figure 12.7   Flow diagram.

needs to be done in order to correct the causes of our problem. The key is not which particular tool is used, but rather that we begin looking for answers from a wide perspective. More specifically, we need to grasp the actual conditions of the work we have identified as the problem point. If our analysis brings out the issues that need correction, we can move on to step 3, which would be to develop a new and improved method that corrects the problem. However, more often than not, what we learn from this overall analysis is not necessarily what specifically needs to be corrected, but only the location of a specific job or situation that is causing the trouble.

Prepare for Solution of Mechanical Problems: Specific Situation Analysis

Once we have located the specific job or situation causing the trouble, we then move on to analyze this using the Job Methods Improvement plan. Here we take on the first two steps of the JM method, which are to (1) break down the job and (2) question every detail. As we learn in the JM training course, the breakdown is a complete and accurate record of the operation made right on the job as it is actually being done and just as you see it, not as you remember it or as you think it should be done. It lists all the details of the job, every single thing that happens, including all material handling, hand work, machine work, inspections, and delays.

Once we have listed all the details, we then move on to question each and every detail using the five W's and one H:

1. *Why* is it necessary? This is the most important question and yet oftentimes the hardest to get answered. If the detail is not necessary, then it can be eliminated.

2. *What* is its purpose? If we're having difficulty answering the *why*, this is a check question to determine if the detail has a useful purpose or adds value. If the detail is found to be necessary, then we continue on with the other questions.

3. *Where* should it be done? Where is the best location to do the job? Why is it done there? Where else could it be done? Can it be combined with another detail?

4. *When* should it be done? When is the best time to do each detail? Must it follow or precede another detail? Are the details in proper sequence? Can it be done simultaneously with another detail?

5. *Who* is the best qualified to do it? Who is best qualified in terms of skill, experience, physical strength, and availability?

6. *How* is the best way to do it? Can it be done easier and safer? Can the layout of the workstation be improved? Are proper tools and equipment being used?

The key to this process is that the answers to these questions, asked energetically and thoughtfully, are the ideas that will lead to improvements and solutions to the problem at hand. At this stage, we only write down these ideas on our breakdown sheet and do not take any action yet; that will be taken up in the next step when we correct the problem. We are looking for new ideas on how to do the work that remove the problem points that are causing the problem. By questioning the details of the work method around that problem point, we can come up with answers to solve the problem.

The Problem Solving course added one more level of analysis here that is not part of the JM method. That is, to look at the three parts of any job:

1. *Make ready.* This is the time and effort spent in getting things ready, such as materials, tools, equipment, etc. Also, the placing of materials or parts in the nearby work areas, from carts, containers, racks, and so on. It includes Total Productive Maintenance (TPM) checks, 5S activities, and safety precautions.
2. *Do.* This is the work that actually does accomplish the desired objective and adds value to the final product or service.
3. *Put away.* This includes all the details necessary to complete the job after the *do* operation is complete. It includes setting the part aside or placing it on carts, in containers, on racks, on conveyors, etc. It would also include paperwork and replacing tools.

When questioning the details of the job, we should first determine which of the three types of work each detail consists of, and then question the *do* details first. If they are found to be unnecessary, then there is no need to question the rest of the operation around that unnecessary detail. If the *do* detail is necessary, then we continue to question the *make ready* and *put away* details. In fact, the greatest opportunity for improvement is in the *make ready* and *put away* details because they add to the time and cost but not to the value of the product or service being created. In most jobs, less than 50% of the total time to get them done is taken up by the *do* part of the job, and so these non-value-added details should be questioned with improvements in mind.

To sum up Step 2 for solving mechanical problems, we should:

1. Analyze the *overall situation* using tools such as the flowchart or flow diagram.
2. Make a job breakdown of the *specific job* where the trouble is located.
3. Question the details to find ideas upon which we can make the needed correction.

Prepare for Solution of People Problems: Don't Know/Can't Do

While some problems are mechanical in nature, the truth of the matter is that almost all of them are, in some way or another, people problems. We said that we can break down these people problems into four distinct types: people who *don't know* or don't understand the work assigned to them, those who *can't do* the work due to lack of skill or practice, people who *don't care* about the work and thus are not motivated to do the work properly, and those who refuse to follow instructions and *won't do* the work as assigned.

We could think of many reasons for why these people problems occur. On careful consideration, though, we can boil them down to three main causes: faulty instruction, incorrect assignment, or personality situations that interfere with the harmony of a good work environment.

Here let's take up the first two categories of *don't know* and *can't do*, which are outcomes due to faulty instruction. What do we mean by faulty instruction? We can categorize faulty instruction into four categories:

1. Insufficient instruction
2. Incorrect instruction
3. Inefficient instruction
4. No instruction

The next question, then, is why do these faulty instruction situations occur? The major cause of these poor forms of instruction would be that little or no preparation has been done of the instructor, the workplace, or the learner. And these are just the points taken up in the preparation portion of the Job Instruction method, which we call Get Ready to Instruct. There are four things we do before we instruct:

1. Make a timetable for training
2. Break down the job
3. Get everything ready
4. Arrange the worksite

Here again we will not go into all the details of the method, which can be found in *The TWI Workbook*. Briefly, when making a training timetable, we determine *who* to train *on which job* and *by when*. The *who* and *on which job* factors are found in Step 1, Isolate the Problem, where we identified the problem point causing the problem. What is left is to make a plan to determine *when* we will do the instruction. In addition, we also consider if it should be broken up into smaller instructional units for ease of training and effectiveness and, if so, look at the order in which we will instruct these smaller units.

Once we have made a plan for training, next we break down the job so that we have notes to clearly organize the job in our minds when teaching. The main reason for poor or incomplete instruction is that the person training the job did not take the time to identify the critical factors in the job, and the ensuing confusion leads to mistakes, rejects, accidents, delays, tool and equipment damage, material wastage, loss of productivity, reduction in morale, etc. The job breakdown allows you to put over the job to the learner quickly and completely because you know just what information you need, in the right amount, and in the correct order. The content of the breakdown includes important steps (what is done), key points (how it is done), and reasons for key points (why it is done that way).

Making breakdowns for Job Instruction is the most critical part of the method and takes great skill and practice. In particular, when we teach the key points, we can be sure that operators know how to do the jobs correctly, safely, and easily. These key points are things that make or break the job (it has to be done just that way), things that injure the worker (safety), and things that make the work easier to do (a knack or feel for the job, a tip or a special piece of timing). These breakdowns don't try to cover every single thing that happens in a job, but are just simple, commonsense reminders of what is key to performing the job.

Now that the instructor is prepared to teach, the next thing is to prepare the workplace. We should have everything ready, like tools, materials, equipment, supplies, and anything else necessary to teaching the job. If we forget something and make excuses, or use makeshift tools and parts, we set a low standard for the work and lose the respect of the worker. We also need to have the workplace properly arranged, just as we would want it maintained in actual working conditions. Otherwise, people will fall into bad habits and accidents are more likely to occur. In other words, we need to set the right example because these first impressions of the job they are learning will carry on into the future and affect the results we get moving forward.

The final thing we need to prepare for solution in these types of people problems is to prepare the learner. This is actually Step 1 of the Job Instruction method, and it gets the learner in the proper frame of mind to be ready to learn. Here we:

- Put the person at ease
- State the job
- Find out what the person already knows
- Get the person interested in learning the job
- Explain tools, equipment, and safety gear (this is an additional item put into the JI method with the Problem Solving course)
- Place the person in the correct position

To sum up, in order to prepare to instruct we should:

1. Prepare yourself (the instructor)
2. Prepare the workplace
3. Prepare the learner

Prepare for Solution of People Problems: Don't Care/Won't Do

Problems concerning people who don't care or won't do are situations concerning attitude and behavior that are unwieldy and difficult to handle. The Job Relations module of TWI was set up specifically to handle these kinds of problems and, more importantly, to show supervisors ways to prevent them from happening or, when they do occur, how to get into them early, while they are still small and easy to resolve. In fact, the four-step method of JR is appropriately titled "How to handle a problem."

So it should not be surprising to see how the JR method, just as it was taught in the original course, fits in perfectly with the problem-solving program. In order to prepare for solution, first we *get the facts*, which means we have to *be sure we have the whole story. Reviewing the record* of the case is a way of being sure we know the full background of how the problem got to its present point. And since this is a problem with people, there are sure to be *rules and customs that apply.* Rules are conditions governing conduct and behavior and are written down in a policy manual that supervisors need to know and abide by. Customs, on the other hand, are precedents or unwritten rules, and many times these are even stronger than the written rules. Supervisors must take these into consideration as well.

Where supervisors usually fall down, though, in failing to prepare for a solution to a people problem, is that they don't *talk with the individuals concerned* in order to *get their opinions and feelings.* This is key to getting the facts because what a person thinks or feels, whether right or wrong, is a fact to that person and must be treated as such. If we fail to hear him or her out and learn where that bad behavior is coming from, then we'll never be able to take the correct action that solves the problem. We'll only be acting on assumptions, usually wrong ones, and attacking symptoms that do not get to the root cause of the problem.

After we get the facts, the next thing to do is to *weigh and decide,* in other words, to analyze the facts and evaluate options so that we can decide on a course of action. This analysis would include *fitting the facts together* while looking for gaps or contradictions in the facts. As we see the connections being made, we then *consider their bearings on each other,* looking for cause-and-effect relationships that help us understand the situation. Based on this understanding, we then consider *what possible actions there are,* being sure that these *conform to practices and policies.* We are now ready to evaluate these options, considering whether each option *meets our objective* and *what effect it will have on the individual, the group, and production.* Through this careful consideration of the facts, we can be sure that we *don't jump to conclusions* and take a rash action, so common when it comes to dealing with people problems, that ultimately ends up not solving the problem.

The above items in italics come directly from the JR four-step method. The Problem Solving course added a few additional points to consider, which only further enhanced the human aspect of the program. While evaluating the possible actions, it advised supervisors to consider:

- What will the results of the action be?
- Question the psychological effect
- Don't hurt the person's pride
- Leave a way open for the individual to save face

Step 3: Correct the Problem

As we saw in Step 2, the contents in Step 3, Correct the Problem, come directly from the four-step methods of JM, JI, and JR. We'll briefly summarize the highlights of each method, but for a full description of how each is used, refer to *The TWI Workbook*. The contents of Step 3 are seen in Figure 12.8.

Develop the New Method and Put Correction into Effect

For mechanical problems, we can make corrections when we improve the method in such a way that the causes of the problem are removed. We can make improvements when the details of the method are eliminated, combined, rearranged, or simplified. When we *eliminate unnecessary details*, we remove waste in our usage of manpower, machines, tools, materials, time, etc. We *combine details when practical* in order to improve problems with transportation, inspections, delays, and storage. We *rearrange details for better sequence* when the problem is one concerning location or excessive handlings, backtracking, delays, accident hazards, maintenance

Mechanical Problem	People Problem	
Develop the New Method	**Present the Operation**	**Take Action**
1. Eliminate unnecessary details	Tell, show, and illustrate one important step at a time	Are you going to handle this yourself?
2. Combine details when practical	Do it again stressing key points	Do you need help handling?
3. Rearrange details for better sequence	Do it again stating reasons for key points	Should you refer this to your supervisor?
4. Simplify all necessary details	Instruct clearly, completely, and patiently, but don't give them more information than they can master at one time	Watch the timing of your action
5. Work out your ideas with others	**Try-Out Performance**	Explain and get agreement on action
6. Write up the proposed new method	Have the person do the job—correct errors	Take the action
Flowchart	Have the person explain each important step to you as they do the job again	Consider the person's feelings and attitude
Job breakdown sheet	Have the person explain each key point to you as they do the job again	Inform everyone involved
Apply the New Method	Have the person explain reasons for key points to you as they do the job again	Don't pass the buck
1. Sell your proposal to the boss	Make sure the person understands	
2. Sell the new method to the operators	**Follow-Up**	
3. Get final approval of all concerned on safety, quality, quantity, cost, etc.	Stress quality and safety	
4. Put the new method to work; use it until a better way is developed	Designate who the person goes to for help	
5. Give credit where credit is due	Encourage questions	

Figure 12.8 Step 3: Correct the problem.

possibilities, or working conditions. And we *simplify all necessary details* when the resolution of the problem lies in reducing nonproductive work motions or making the work easier and safer to do. We are able to make these kinds of improvements based on the ideas we found in Step 2, Prepare for Solution.

In developing the new method, it is critical to write it down in the form of a proposal. This written proposal, which includes explanations of the current and the proposed methods, as well as the expected results, will be indispensable when we apply the new method, in other words, when we *sell the proposal to our boss* and *get final approvals of all concerned on safety, quality, quantity, cost, etc.* It will be easy for these other parties to recognize and approve our changes when they know where our ideas came from and how they will go about solving the problem at hand. In addition, we must also make an effort to *sell the new method to the operators* because, without their cooperation, the changes will not be implemented and the problem will continue. The best way of getting this cooperation is to involve the operators right from the start of the improvement process, using their ideas, along with any others, and *giving credit when due.* Finally, we must not let waiting kill our ideas and *put the new method to work* right away.

Instruct the Learner

For problems concerning people who don't know or can't do, in Step 2 we prepared ourselves, the workplace, and the learner for instruction. It is our task in Step 3 to deliver that instruction and teach the learners how to do the job correctly, safely, and conscientiously in order eliminate problems that result in scrap, rework, delays, accidents, damaged tools and equipment, spoilage of materials, low productivity, etc. We *present the operation*, telling what you are showing, one important step at a time, following the job breakdown sheet we made in Step 2. Then we do the job again, this time stressing the key points for each step, again following the breakdown sheet. We do it yet one more time, explaining the reasons for the key points. We can repeat this as often as necessary until they fully understand the job.

Once they understand how it is done, then have them do it in a *try-out performance*. The first time they perform the job, be sure to correct any errors so they don't develop bad habits—we do not want them to learn from their mistakes, but to repeat the correct performance that was presented to them by the instructor. Once you are sure they are able to do the job correctly, have them do it again, telling you the important steps as they show you the job step by step. Have them do it again, this time stressing the key points. Have them state the reasons for each key point as they do it a fourth time. You can repeat this process until you are completely confident that they understand the job fully.

At this point, you can put them on their own, stressing the need to pay attention to their new work in terms of quality and safety. Be sure to encourage them to ask for help when needed. They are now able to do the job, but they are not yet experienced at it, and problems and questions will certainly come up. People are almost always hesitant to ask for help because they see it, mistakenly, as a sign of weakness and are afraid of negative consequences should they show their ignorance. Let them know that you are ready and willing to help when needed, and assign a backup person in case you are not around.

Many decades of experience has shown that following this method of instruction produces good results and prevents many, if not most, of the common day-to-day problems supervisors have to deal with. In fact, a high percentage of production problems can be traced back to poor training, and the JI method is a powerful means of solving these problems.

Take Action on the Problem

For problems concerning people who don't care or won't do, these problems require quick and careful solution. In Step 2, we collected the facts, weighed them carefully, and came to a decision on how to fix the problem. Now, in Step 3 we have to *take action* on the problem and make the correction.

The important things to remember here are:

- Are you going to handle this yourself? Is it your *responsibility* to deal with this problem, or should you pass it on to someone else?
- Do you need help handling it? Do you have the *ability* to take this action, or do you need help from human resources, legal, professional counselors, etc.?
- Should you refer this to your supervisor? If you have the *authority* to take the action, don't bother your boss.
- Watch the timing of your action. What is the best time and place to take the action?

The problem-solving manual added a few more points to consider, up and above the method taught in the JR program. Here again, they were very concerned with the human approach to problem solving and stressed the overarching need, as dictated in the JR method, of maintaining strong relations with the people even as you go about taking action on problems that concern them:

- Explain the action to the person—why it is best for him or her.
- Give advantages and benefits; get acceptance.
- Consider the person's feelings and attitudes when you take the action.
- Secure clear understanding.
- Inform everyone involved.

Because dealing with people problems is uncomfortable and unpleasant, it is very tempting to pass them on to others, like your boss or the HR department. When we do that, though, we lose standing with our people, and that propels the downward spiral of deteriorating human relations. When taking action on problems, we must *never pass the buck*.

Step 4: Check and Evaluate Results

Once the correction of the problem has been made, it is vital to ensure that it takes hold and sticks and that the problem does not reoccur. The contents of this step are as follows:

- Follow up to see that the change or correction has been made.
- What improvement do the records show in quality, quantity, safety, and cost?
- Consider the human angle. Note changes in attitudes and relationships.
- Inform all those concerned of the progress and results of the action or correction.
- Look for ways to prevent a recurrence of the problem.

As one would expect, the key here is to find and evaluate data that show that the correction has worked, and to do this we can refer to records on production, quality, cost, safety, productivity, attendance, grievances, and so on. The method also stresses that, once we have this information, we inform all those concerned on the progress being made and the results of the correction. This is particularly important when we consider the fact that sometimes our actions to correct problems may actually create new problems. By keeping a close check on this possibility and keeping everyone informed of the changes made and results obtained, we can be sure that the problem has been solved in a way that satisfies all parties.

As we have seen throughout this process, and in the overall TWI philosophy, the human angle is always given special attention, and even more so here because whenever changes are made, people will inevitably build up resistance or resentment to the change. These kinds of negative feelings will present themselves in the attitudes, behavior, and relationship situations of the people—not only the people directly involved with the problem, but also the overall group that will be affected by the change. It is simply a part of human nature that people react to change as a threat to their basic needs surrounding their jobs: stability, recognition, security, opportunity, participation, job satisfaction, etc.

When people resist change, it is an emotional reaction based on fear, not reason. They fear that their familiar routines will be upset, forcing a change in the work habits they have become accustomed to. They worry if they will be able to learn the new methods and what effect these new methods will have on their output and their working conditions. When making changes to correct problems, then, telling people in advance about those changes will help pave the way for acceptance. In particular, telling them why it necessary in a clear, simple, and concise manner while showing the benefits of the change to everyone involved helps to take down the resistance. If nothing else, just giving them a chance to "blow off steam" and have some say in the matter will help them to agree to the new standards.

Oftentimes people interpret our effort to solve problems as a personal criticism. In other words, they see the change as a way of blaming them for what has been done in the past and up to now. People do not like being told they are wrong, so in discussing the problem correction we have to take special care to refrain from criticism and make it clear that we are looking for constructive actions on solving problems. Take special care, as well, when dealing with the people who originally proposed or installed the current methods so they do not feel the need to defend what they have done in the past.

Summary

Because the TWI Problem Solving method contains the full depth of all three of the other TWI methods, it is a much bigger and robust program. To properly use it, then, a supervisor must be well versed in Job Methods Improvement, Job Instruction, and Job Relations skills. The practice of these three essential skills *is* the practice of problem solving, with the addition of the all-important Step 1, isolating the problem—finding the true root cause first. Once we have identified these root causes, or problem points, we can quickly and effectively use our TWI skills to bring the problem to resolution.

In this way, the TWI approach to problem solving gives renewed meaning to the TWI skills and allows front-line supervisors to solve their everyday problems using the TWI skills they already possess. It also highlights the true leadership potential available to supervisors when they master these essential skills. A great leader is one who can guide the team to its objectives by overcoming the many obstacles that get in the way. These obstacles are problems, and problem solvers are the people we will follow because they have the skills to lead us to our goals.

Conclusion

A New Beginning for TWI

We were excited about the opportunity given to us by Productivity Press to put together a manuscript that became *The TWI Workbook: Essential Skills for Supervisors* in 2006. Patrick had long dreamed of capturing his knowledge of TWI, much of which was passed on to him over years of mentoring by his sensei Kazuhiko Shibuya. Bob, seeing the book as a way to establish TWI as basic training for supervisors in a lean environment, was certain organizations would benefit by learning how to properly utilize each TWI skill to produce needed results as was done during WWII.

Thanks in big part to the publication of *The TWI Workbook*, along with the other books describing *The Toyota Way* that we have cited throughout this work, usage of TWI has increased and we feel very privileged to have had the opportunity of working with so many companies that have benefitted as a result. The case studies presented in this book are only a few of the increasing number of organizations that are energetically applying these timeless methods and obtaining great results. The list of these organizations continues to grow day by day.

In writing *The TWI Workbook* that explained the 4-Step Method for each of the TWI skills of JI, JM and JR, we methodically went through a step by step process to present the instructions and exercises that, if followed, a person could learn to perform the skills with some degree of proficiency. However, as we saw in this work on *Implementing TWI*, the methods, though seemingly simplistic on the surface, are full of deep wisdom that can only be learned through coaching, practice, and experience. This wisdom, unfortunately, cannot be boiled down to a list of bullet points or a flow chart. It has to be internalized and "lived"—in other words, it has to find its way into the culture of an organization so that people literally "live and breathe" it every day. As we saw in all the case studies in this book, that process is not a "procedure" but a "way of life."

In these chapters, then, we have endeavored to describe, in our explanations and through the cases study examples, the philosophy of TWI as much as the practice. For if you cannot capture the spirit of how these programs lift people up and carry them to new heights, you will have missed the true value of what has made this training so effective over many generations of leadership.

Companies that embraced TWI early on did not get full benefit when treating TWI modules as just another tool of Lean. This was before the true connection between TWI and the Toyota Production System was made known. We ourselves learned much from these companies and from other books increasing our TWI implementation learning curve working with companies that struggled to make TWI a part of their culture. That is when we started gathering data from these many organizations that came to TWI as a solution to production problems they could trace back to a lack of standardized work. Their leaders had already progressed with Lean-Six Sigma to a point where, in search of a solution to standardized work, they began investigated JI on their own once they saw the connection. It took a few years for these companies to progress to the point where we could compile the case studies in this book as examples for others to follow, and we are deeply indebted to them for sharing their stories.

We hope that you have benefited from reading this book and share what you learn with others who are, no doubt, experiencing the same "trials and tribulations" you have gone through in your continual search for "a better way." For those of you that either are or will soon be challenged to replace experienced employees as they retire, the search for a better way begins now by capturing the knowledge these people acquired on the job before they exit the workforce. When treated with respect people are more than willing to share what they have learned on their own: a "knack", a "trick", or a better way of doing a job so that others do not have to relive past mistakes when learning to do their jobs.

In his book *Management*, Peter Drucker summed up our leaning experience with TWI over the past few years as follows. "The first step toward making the worker achieving is to make work productive. The more we understand what the work itself demands, the more can we then integrate the work into the human activity we call working. The more we understand work itself, the more freedom we can give the worker. There is no contradiction between scientific management, that is, the rational and impersonal approach to work, and the achieving worker."* As the case studies in this book make clear, we think Mr. Drucker would agree with our conclusion that TWI bridges the gap between the impersonal approach to work and the achieving worker to the benefit of the entire organization.

---

* Peter Drucker, *Management: Tasks, Responsibilities, Practices,* Harper & Row, Publishers, New York , 1973, p. 199

Index

About the Authors

Patrick Graupp began his training career at the SANYO Electric Corporate Training Center in Japan after graduating with highest honors from Drexel University in 1980. There he learned to deliver TWI and other training programs to prepare employees for assignment outside of Japan. He in turn was also transferred to a compact disc fabrication plant in Indiana, where he obtained manufacturing experience before returning to Japan to lead Sanyo's global training effort. Patrick earned an MBA from Boston University during this time, and he was later promoted to the head of human resources for SANYO North America Corp. in San Diego, where he settled.

Working with Bob Wrona, Patrick took vacation time in 2001 to deliver a pilot project for CNYTDO, the predecessor and parent company of the TWI Institute, to reintroduce TWI into the United States. The results encouraged Patrick to leave SANYO in 2002 to deliver and spread the TWI program as he was taught in Japan and which he described in his book *The TWI Workbook: Essential Skills for Supervisors*, a Shingo Research and Professional Publication Prize Recipient for 2007. Since then he has developed hundreds of trainers who are now delivering TWI classes across the country and around the world.

Robert J. Wrona began his manufacturing career at Chevrolet in Buffalo, New York, where he was promoted to supervisor after earning his BS from Canisius College. He moved on to Kodak in Rochester, New York, where he became interested in organizational development while earning his MBA from Rochester Institute of Technology. Bob joined a high-volume retail drugstore chain in Syracuse, New York, when it was a twelve-store operation. He standardized store operating systems and procedures, developed internal training, and reorganized central distribution as the company profitably grew into a regional chain of 140 stores in eleven years.

Not content with becoming an administrator, Bob returned to manufacturing as an independent TQM consultant for small manufacturers to engage their people to improve performance. Fifteen years of hands-on implementation made it clear that supervisors lacked the skills to lead in the new world of Lean manufacturing. He discovered TWI when studying kaizen and tracked down TWI master trainer Patrick Graupp in 1998. The opportunity for them to reintroduce TWI in the United States came in 2001, when Bob became a Lean consultant for CNYTDO, Inc., which provided support for reintroducing TWI in Syracuse, New York, as detailed in his 2007 Shingo Prize-winning book *The TWI Workbook: Essential Skills for Supervisors*.

Printed in the United States
by Baker & Taylor Publisher Services

Business Improvement / Lean

"Graupp and Wrona have been teaching and implementing the tools of TWI for years in many different types of companies … . If you want to get from interesting displays to true standardized work, read this book."
—Jeffrey K. Liker, author, *The Toyota Way*

"… explains why TWI is a basic building block for converting an organization that has engineered Lean into a learning organization … . uses cases to explain how to create no-nonsense culture change by teaching people how to do work differently, and how to relate to each other differently in order to work more effectively."
—Robert "Doc" Hall, Editor-in Chief. *Target Magazine*, author of *Compression: Meeting the Challenges of Sustainability Through Vigorous Learning Enterprises*

"Graupp and Wrona … not only explain the lessons learned … but further clarify the integration needed to develop a strategic success between TWI and other core functions of any business."
—Jim Huntzinger, Founder and President, *Lean Accounting Summit, TWI Summit,* and *Lean and Green Summit*

"If companies are serious about developing skills and making improvements, then I urge them to study up on TWI concepts."
—Art Smalley, President, Art of Lean, Shingo Prize Research Award winner, 2003

"… Graupp and Wrona bring many examples of companies that [improved] competitiveness by improving their capacity to fully engage their workforce … ."
—From the Foreword by Steven Spear, author and *Sr. Lecturer, MIT Sloan School of Management*

Implementing TWI: Creating and Managing a Skills-Based Culture

Featuring strategies employed in Lean, this volume describes the experiences of organizations using TWI more than 60 years after the Training Within Industry program turned the U.S. into the industrial giant that won World War II. Based on their experience implementing TWI in organizations as diverse as Virginia Mason Medical Center and Donnelly Manufacturing, Shingo Prize Winners Patrick Graupp and Robert Wrona prove why many consider them the most successful TWI trainers in the world. Their hands-on manual provides the tools and templates that can turn any company's employees into a skilled invested workforce capable of realizing unprecedented profits.

CRC Press
Taylor & Francis Group
an **informa** business
www.crcpress.com

6000 Broken Sound Parkway, NW
Suite 300, Boca Raton, FL 33487
270 Madison Avenue
New York, NY 10016
2 Park Square, Milton Park
Abingdon, Oxon OX14 4RN, UK

K11246
ISBN: 978-1-4398-2596-9

90000

www.productivitypress.com